Scientific Visualisation

Marianne Ina Richter

Scientific Visualisation

Epistemic Weight and Surpluses

PETER LANG
EDITION

Bibliographic Information published by the Deutsche Nationalbibliothek
The Deutsche Nationalbibliothek lists this publication in the Deutsche Nationalbibliografie; detailed bibliographic data is available in the internet at http://dnb.d-nb.de.

Zugl.: Stuttgart, Univ., Diss., 2013

Library of Congress Cataloging-in-Publication Data
Richter, Marianne Ina, 1984-
Scientific visualisation : epistemic weight and surpluses / Marianne Ina Richter. - 1 [edition].
pages cm
ISBN 978-3-631-64300-6
1. Research–Methodology. 2. Creative thinking. 3. Visualization. 4. Science–Methodology. I. Title.
Q180.55.M4R53 2014
001.4'2–dc23 2013044745

Cover illustration:
© Martin Falk

D 93
ISBN 978-3-631-64300-6 (Print)
E-ISBN 978-3-653-03166-9 (E-Book)
DOI 10.3726/978-3-653-03166-9

Peter Lang Edition is an Imprint of Peter Lang GmbH.

Peter Lang – Frankfurt am Main · Bern · Bruxelles · New York · Oxford · Warszawa · Wien

www.peterlang.com

Table of Contents

Abbreviations and Acronyms

C	Chondrogenic cell (Chondrocyte)
DN model	Deductive-Nomological model
Dof	Depth of field
MAPK	Mitogen-Activated Protein Kinase
MAPKp	Mitogen-Activated-phosphorylated Protein Kinase
MB	Megabyte
MSC	Mesenchymal Stem Cell
O	Osteogenic cell (Osteoblast)
P	Progenitor cell (Progenitor)
PI	*Philosophical Investigations* Wittgenstein (2001) [1]1953
TLP	*Tractatus logico-philosophicus* Wittgenstein (1974) [1]1922
TR	Transcriptional Regulator
VB	Vertex Buffers
x_C	Chondrogenic TR
x_O	Osteogenic TR
x_P	Progenitor maintenance factor
z_C	Chondrogenic stimulus
z_D	Pro-differentiation stimulus
z_O	Osteogenic stimulus
2D	two-dimensional
3D	three-dimensional

Table of Figures

Copyrights

List of Tables

Preface

The philosophical interest in scientific visualisation had already reached a peak when I started working on this thesis in 2009. Nevertheless, the various contributors to the thriving debates have kept on nurturing diverging intuitions on the nature, exploitability and methodological status of visual – and especially *pictorial* – means. It seems that while the prevalence of pictorial means was never called into question their status as aids to the scientist – and thus as subjects of interest to the philosophy of science – is repeatedly in need of reconsideration. Meanwhile, the range of attributions comprises everything from *redundant* or *distractive* to *enabling* or *instructive*, which only reinforces the friction between "those who think that the image is an extremely rudimentary system [...] and those who think that signification cannot exhaust the image's ineffable richness"[1] (Barthes 1977, p. 32). The implementation of study programs, such as *Visual Computing* which grew out of computer science, or *Visual Studies* which involves several disciplines within the humanities, added further fuel to a fire which finally spread to the philosophy of science as well, where it presents a particluar challenge to those who try to analyse the notion of science (and notions related to it) in terms of its integral and contingent components. Following this track, the question arises as to whether the long-neglected pictorial means can be integral to core measures of scientific practice and thus carry epistemic weight in a non-trivial sense.

In my thesis I have also made attempts to use this strategy of conceptualisation. That is, I assume that there are measures, such as representation and argumentation, which are integral to the scientific practices under consideration, and which can be modelled in such a way that we can truly say whether or not a pictorial symbol system has the capacity to become integral to *them*. Needless to say, I am not about to resolve once and for all the tensions between different ideas and intuitions that such an analysis brings about with regards to the role and performance of pictorial means in science, even more so as these tensions arise in various empirical

1 Translation by S. Heath (Barthes 1977) of the following quote: „les uns pensent que l'image est un système très rudimentaire [...], et les autres que l' signification ne peut épuiser la richesse ineffable de l'image“ (Barthes 1964, p. 40).

contexts, such as research-, conference- and publishing-practices, which is why one can hardly uphold the claim that they are principally or broadly induced by a certain, perhaps incorrect running argument or inadequate framework. Instead of tackling an over-generalised idea of the so-called 'state of art' and its deficiencies, I therefore decided to spell out the traits of exemplary scientific practices in order to obtain provocations and correctives for the discussion of – again *exemplary* – epistemological and aesthetic accounts. The choice of reference frameworks is substantiated by two literature surveys – one with a methodological and one with a conceptual focus. Provided the reader (philosopher, scientists, both or other) identifies with the chosen frameworks and revisions that I would like to propose to the recent discussion, s/he will gain a conceptual toolset that helps at least to take a local stance on the controversy.

My work on that toolset would not have been possible without the vital support that I received at the University of Stuttgart and beyond. At this point, I would like to thank all persons and institutions that helped me to complete this project. First of all, I would like to thank my examiners Jun.-Prof. Dr. Ulrike Pompe-Alama, Prof. Dr. Thomas Ertl and Prof. Roman Frigg, Ph. D., for their encouragement and guidance throughout this project. I also owe gratitude to Prof. Dr. Christoph Hubig who followed the project with vigorous interest and who was the examinder who passed it for the *Milestone Examination*. For their valuable advice and time devoted to discussing my ideas, I am also grateful to Prof. Dr. Catrin Misselhorn, Prof. Dr. Dr. Rafaela Hillerbrand, Angela Matthies, M. A., Dr. Juan Duran, Dr. Eckhart Arnold and Dr. Tillmann Pross. Furthermore, I would like to thank Michael Poznic, M. A. and Dipl. Musician Tanya Newman for proofreading the first drafts. Special thanks go to my colleagues from the natural and engineering sciences, especially Dipl.-Inf. Daniella Schittler, Dr. Martin Falk, Dipl.-Phys. Michael Raschke and Dipl.-Phys. David Molnar, who allowed me to participate in their work and who incited me to do so by their curiosity about a 'philosophical view'. A great deal of interest has also been shared by the Richters, especially Sabine, Edda and Rudolf, and my friends, whom I cannot thank enough for following me through the highs and lows that accompanied this research adventure. Last but not least, I would like to thank the Cluster of Excellence Simulation Technology (EXC SimTech) and the Deutsche Forschungsgemeinschaft (DFG) for making it all possible.

Stuttgart, October 2013
Marianne Richter

All sciences must deal with the problem of selecting and constituting "working objects"
Daston & Galison (2007), p. 19

I Introduction

Despite their prevalence, visual – and especially *picture-like* or *pictorial* – measures have been afforded only sparse attention in the philosophy of science. This is perplexing as practice indicates that the ever-present 'scientific figures' (abbreviated 'Fig.') are conceived of as ingredients of both pictorial and argumentative systems. Fig. 1, for instance, exemplifies a strategy of visualisation that is meant to render the numerical solution of simulations comparable to the outcomes of confocal laser scanning microscopy. More precisely, Fig. 1 adopts the associated optical effects as displayed by the targeted micrographs (cf. Fig. 2), such as the afterglow of the stimulated fluorescent markers and the stark contrast between the glowing and the darker areas. Hence, the numerical solutions of the simulation – which could actually be presented much more articulately, considering the source matrices on which they are based – are rendered indistinct, up to a degree where it becomes difficult to distinguish between all the distinct values that are used to represent foreground and background, cell tissue and plasma, etc. Still, on account of these very figures, it is argued that there are differences between the diffusion of signalling molecules and their transport via motor proteins, such as crowding behaviour in the case of transport (Fig. 1 [right]; cf. Falk et al. 2009, p. 173). It thus seems, after all, that these figures are truly used to provide support for arguments. In the above case, they give credence to the claim that there are different modes of cellular signal transduction, which is true insofar as the signalling molecules crumble in one setting whilst they do not crumble in another, and insofar as the terms 'crumbling' and 'molecule' express what is primarily expressed by means of the figures. In other cases, one might argue (just as cogently) that this is true in reference to a set of equations, which would then share the same methodological status. In order to thus admit the epistemic weight of pictorial means, what matters, ultimately, is that the use of these figures fits not only the conditions of dense, picture-like systems, but also distinct, argumentative, ones.

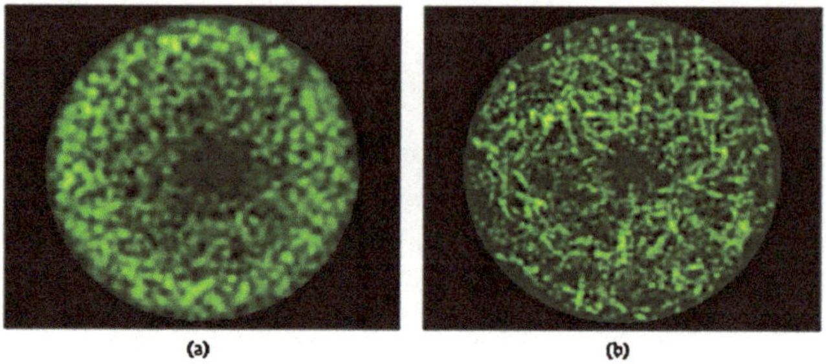

Fig. 1: *Microscopy*-like *images (cf. Falk et al. 2009, p. 173)*

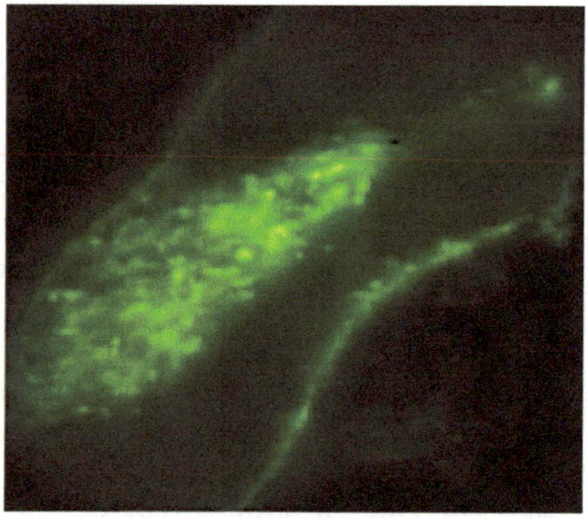

Fig. 2: *Confocal laser scanning micrograph (cf. Falk et al. 2009, p. 169)*

But is it right to assume that this is the case? Can pictorial systems carry epistemic weight in such a rigid sense? Can they even bear any surpluses, compared to other semiotic types? In view of practical examples, such as the one presented above, recent literature has delved into the conceptual framing of dedicated beliefs.

Many contributors, however, tend to focus on one of two cases. Either they focus on semiotic types that do not pose severe problems of systematic integration, such as graphs or diagrams, which are distinctly composed (resp. approached) right from the beginning and which are thus attuned to the semiotic requirements of arguments (cf. e.g. Perini 2002ff.). Or they focus on picture-aided measures that are pre- or post-preparatory to arguments but not integral to them in the strictest possible sense (cf. e.g. Lynch 1988). Yet other authors reject the mission completely, arguing that there cannot be genuine picture-like (resp. analogue) systems on a discrete computer basis (cf. Fischer 1997), which would render the problem of the epistemic weight of pictorial systems obsolete or trivial. In any case, recent literature gives up rather quickly and thus falls short of the ambition to understand the integration, role and performance of genuinely pictorial means and measures in formally established, discursive systems.

This book seeks to tackle the deficits of the present discussion. The reader can thus expect a review of arguments on the state and status of pictorial means at the permeable interface between scientific research and communication as well as a counterpoise to approaches that either deny attention to pictorial means within that scope, or – and this is still the common mode in the philosophy of science – ones that pigeonhole them into the fuzziest category available (e.g. into the group of models or cognitive auxiliaries, or both). The central claim is, indeed, that pictorial means carry epistemic weight in the suggested sense of being integral to arguments (cf. also Perini 2002ff.), which can be explained when taking into account the as yet missing concept of *ad hoc syntaxes*. Ad hoc syntaxes explain why symbol systems may be picture-like, formally tractable and efficient at the same time. The aim of this book is to distil and make clear this set of conclusions.

This book builds on a dissertation and preserves its style and structure for the most part, thus offering different options for taking in its contents. Readers who are interested in the chosen approach towards pictorial means in science may be interested in the methodological framing of the presented philosophical endeavour. In this case, they should proceed chronologically, beginning with the next chapter which starts off with the clarification of the chosen terminology (cf. 2.1) and the assumed philosophical problems (cf. 2.2), and which motivates the attempt to solve these problems with reference to the history of picture-aided science and the devoted, philosophically established or affined, mind-sets (cf. 2.3). Someone seeking for a more profound idea of the targeted scientific practices should perhaps skip these parts and jump directly to the introduction of examples, some of which have not been documented in the philosophy of science so far (cf. 3.1 and 3.2). The study of these examples has profited greatly from the exchange with local scientific staff and from the consultation of their papers and lectures, and it is meant to help acquire and fortify intuitions about the nature, role and performance of pictorial means in the targeted contexts. In this spirit, the third chapter concludes with the abstraction of findings that set demands on any devoted conceptual approaches (cf. 3.3). The main arguments on the epistemic weight and surpluses of picture-like means are presented in the fourth chapter. This chapter offers another option to start this book – especially for those interested in the position defended here. The chapter starts with a literature overview that pinpoints the deficits of the recent discussion and narrows down the range of reference approaches (cf. 4.1). It then extends the most promising one by the notion of ad hoc syntaxes in order to achieve the integration of the neglected genuine pictorial systems into the scope of epistemically 'weighty' systems (cf. 4.2). Finally, the framework is analysed on its implications for the discussion of deficiency and efficiency arguments (cf. 3.3).

II Picture-like Means in Science

2.1 'Figure' – 'Image' – 'Picture'

What do we refer to as a 'figure', an 'image' or a 'picture'? Apparently, there is no definite answer to this question, but there are as many answers as there are ways of using these terms. Equivocation may, of course, occur with *any* terms, including philosophical terms of art (cf. Gadamer 1972, p. 392). Nevertheless, it seems particularly pronounced among the terms that form this section's title. Why is that? This is difficult to say. Could it be a cause or a consequence of the fact that the so-termed entities often work as *links* between different contexts? For example, they link different arenas of science, such as research programs or projects, where they occur as shared objects of focus:

> "[V]isual culture draws on nearly two dozen fields in the humanities, including history and art history, art criticism, art practice, art education, feminism and woman's studies, queer theory, political economy, postcolonial studies, performance studies, anthropology and visual anthropology, film and media studies, archaeology, architecture and urban planning, visual communication, graphic and book design, advertising and the sociology of art" (Elkins 2003, p. 25).

This is still incomplete since Elkins (2003) skips the natural and engineering sciences in this list. Additionally, the so-termed entities also work as links between distinct stages of research, where they serve as pre- or post-preparatory means or measures:

> "Once the researchers succeeded in visualizing these aspects of the flow [i.e. of an airflow inside a storm], they were able to make use of these visual representations to identify some of the key structures and trajectories in the inner dynamics of the flow of air, ice, and water through the storm system. They were able to use this knowledge to construct 'a model of storm evolution and persistence'" (Winsberg 1999, p. 284).

Apparently, so-called 'figures', 'images' or 'pictures' work as links in different respects. They may, however, be characterised differently depending on which side of the bridge they are to be found. And if the linked contexts grow complex or distant, it is likely that the usage of shared terms cannot easily be kept congruent and functional in all of them. What remains are shared words rather than shared pairs of terms and ideas.

The choice ofkey terms and usages is thus integral to approaching the discussion(s) under review, for it contributes to pinpointing the closeness of related approaches. Defining a terminology – that is, constraints for the use of terms – thus means choosing a position on the research landscape. As mentioned above, this is particularly important in research on visual devices where certain key terms – especially 'figure', 'image' and 'picture' – are used so extensively that the *tertium comparationis* (i.e. the common quality of the denoted entities or acts of denotation) becomes vague, if not blurred (cf. 2.1.1). On the other hand, one cannot discard these terms without losing them as links to valuable approaches; and in order to recreate these links, one would have to comment on these terms anyway. Hence, it seems that there are only a few options left for coping with the variance of terminologies that coins the targeted debates, while the option I chose needs to score well against this background. Therefore, I would like to begin by ruling out different strategies of connecting terms and concepts, in order to narrow down the scope of reference works in the first place. The preferred connections are introduced later on (cf. 2.2.2, 4.2).

2.1.1 Extended definition

First of all, one may consider the diversity of usages as fully appropriate due to the fact that the terms under review are commonly used to denote an exceptionally wide range of entities. W. J. T. Mitchell (1984), for instance, drew attention to the fact that "[w]e speak of pictures, statues, optical illusions, maps, diagrams, dreams, hallucinations, spectacles, projections, poems, patterns, memories, and even ideas as images" (Mitchell 1984, p. 504). In the face of this – perhaps still incomplete – inventory, Mitchell suggested the following: "begin by thinking of images as a far-flung family" (Mitchell 1984, p. 504f.); because, like the members of a human family, the things called 'image' share a common "name" (Mitchell 1984, p. 504) and some similarities (e.g. concerning their outer appearance or social role), but do not share any features that would mark *only* them and *always all* of them. From this perspective, which takes the vagaries of everyday language at face value, the definition of preferences and constraints for the use of the term 'image' (and for others alike) would, indeed, look like a privileging of certain family members, which always runs the risk of being judged reductionist or non-intuitive.

2.1.2 Specifications

On the other hand, one could accept the objections from the 'all-in' strategists in favour of a tentative specification of a term's use, which draws focus on certain instances while retaining the option to remove or change this focus in a later trial. This strategy is actually quite popular amongst scientists, since it saves them from constantly mentioning exceptions all the time, and likewise offers them a range of further possibilities for adapting the use of terms to individual focal points.

2.1.2.1 History-based specification

One of those strategies for adapting terminological preferences and constraints is what may be called the 'history-based or 'history-oriented' strategy. This strategy proposes comprising an inventory of usages (as the one presented by Mitchell 1984; cf. 2.1.1) and pointing out exceptional, early or contemporary usages. The favoured one is then substantiated against this background as being 'standard' or 'current'.

This strategy is employed by Mitchell (1984) himself, who offers a "family tree" (Mitchell 1984, p. 505) – that is, different roots and applications of the term 'image' – and who tries to find out on this basis (among other points) "which members of the family of images are called by that name in a strict, proper, or literal sense, and which kinds involve some extended, figurative, or improper use of the term" (Mitchell 1984, p. 506). Mitchell assumes in this context that there are actually stronger intuitions on what is to be taken as an extended use, namely applying the term 'image' to mental entities, such as dreams, or verbal entities, such as metaphors (cf. Mitchell 1984, p. 507).

Independent from the issue-related arguments and measures that guide his approach, it is already possible here to identify the weak point in this type of terminological specification, whether or not it is meant to be tentative, which it is in this case (cf. Mitchell 1984, p. 507ff.). The weak point seems to be the notion that there is an *original* and therefore conceptually or temporally *prior* use of the term. Apart from the troubles of empirical justification, this idea runs the risk of being suspected of (or mistaken for) some sort of terminological normativism, that is, the wish to enforce one usage at the cost of successful variations. Of course, I suppose Mitchell did not mean to use the attribute 'improper' in any such pejorative sense, but rather to provoke a comparison between what is commonly considered as "images 'proper' and their illegitimate offspring" (Mitchell 1984, p. 507), ultimately arriving at the conclusion that "[r]eal, proper images have more in common with their bastard children than they might like to admit" (Mitchell 1984, p. 507). But still, and not the least due to this outcome, Mitchell's approach should help reveal that it is rather unfortunate to refrain from certain usages of a term for the suggested reason that they are less "proper" (Mitchell 1984, p. 506), "literal" (Mitchell 1984, p. 506) or "obvious" (Mitchell 1984, p. 506), when it maintains at the same time that they all work well – at least there seem no further footnotes needed in order to prove the inventory of usages correct and familiar (cf. Mitchell 1984, p. 504) – and when it also maintains that there are less context-privileging options for justifying terminological constraints.

2.1.2.2 Case-based specification

In search of alternative options of definition, one comes across the 'case-based' mode of specifying the use of terms. This mode amounts to one basic measure: one specifies the intended usage(s) performatively, that is, by applying the respective term to a range of concrete instances, which are quoted or otherwise made present at the very moment when the term is used, for example when it occurs in a text.

A. Beyer and M. Lohoff (2005), for instance, edited a volume of essays on scientific functions of 'images' ("*Bilder*"), where the term 'image' ("*Bild*") applies to each sketch, diagram, graph, etc. that is inspected as an example. By this means, a partial definition is performed, meaning that the editors and authors point to *some* of the instances of the term, or that is rather to say, they use the term *ostensively* – like '*this* is an image' and '*this* also'. The only limiting factors are thus given by the medium for presentation, for example by the size of the volume, and perhaps additionally by some implicit rules of the game, like agreements on what can be termed 'image'. For this reason, the case-based strategy may appear somewhat minimal; yet for the same reason it seems quite popular. In particular historians of art and science, like Beyer and Lohoff (2005), are likely to remain undecided on the exact meaning of the term, as this keeps their approaches sensitive and adaptive to conceptual shifts and changes over history (cf. Elkins 2011, p. 15).

The disadvantage of this strategy is such combinations of terms and instances might collide with others for no specified reason. Some authors like L. Perini (2002ff.) would, for example, refrain from applying the term 'image' to all examples presented by Beyer and Lohoff (2005). Moreover, it is also difficult to extrapolate from one combination of terms and instances to others due to a lack of explicit rules for these combinations. Hence, if nothing else, the case-based mode of specifying the use of terms relies strongly on implicit agreements, internalised rules and canonical points of reference.

2.1.2.3 Exclusive concept-based specification

There might thus be the demand for conceptually justified constraints, which guide the distinction between correct and incorrect usages of a term (whether in a descriptive or normative sense) by explicitly profiling the intended or not intended reference entities and application contexts.

Thanks to such an explicit account of the conditions of use, 'concept-based' strategies have proven suitable in the science-theoretic field; because when it comes to characterising and explaining an entity's epistemic weight or capacity, an analysis of its components and embeddings is required anyway (cf. 2.2.1.2). Hence, these strategies can be smoothly combined with further concept-related work on the instances chosen. This might actually be the reason why there are many versions thereof, also amongst those approaches that focus on visual devices in the scientific field.

K. Hentschel (2000), for instance, introduces the term 'nonverbal means' – suggesting that its instances (including "visual representations", Hentschel 2000) are characterised by anything *except* by the features that are identified as 'verbal'. Accordingly, he characterises the entities termed 'visual representations' indirectly in the way that he points out antonyms and contrary terms of application.

While this is certainly an effective strategy for defining the use of terms, it might not be the most efficient one, since by listing what something is *not*, one offers merely exclusive constraints, like: what is classified as 'visual' cannot be classified as 'verbal', although any other type of classification is still possible. Hence, the advantage of this strategy, which lies in the evocation of conceptual controversies, needs to be balanced with the fact that 'ex-negative definitions' – that is, definitions that clarify what something is *not* – are established on basis of *open lists* of defining features or aspects. For this reason, they work best in combination with others (e.g. with the case-based mode like in Hentschel 2000).

2.1.2.4 Inclusive concept-based specification

However, as indicated before, the scope of concept-based strategies is not restricted to the ex-negative variant. R. Kötter (2007), to name but one of numerous proponents, offers an 'inclusive concept-based' specification, meaning that he aims to capture those features that make the entities termed 'image' distinct from others – and not others distinct from them (which takes endless efforts of specification; cf. 2.1.2.3).

Kötter's particular list of defining aspects can be summed up as follows: An 'image' is an object (1) with a designed ("*gestaltet*") surface that is (2) visible to the unaided eye, (3) created by human beings, using various means, and that (4) is used to represent something (cf. Kötter 2007, p. 164). Accordingly, Kötter espouses the notion that the term 'image' denotes some class of *device*, which is designed for a certain purpose (here: representation) following more or less context-related conditions of success and quality (here: perceptual access). The most prominent and currently uncontested implications of this notion are that humans need to *learn* the handling of 'images' and that they can *fail* on this lesson – either in the sense that they fail on a particular 'image' or in the sense that they fail to gain "visual literacy" at all (cf. Elkins 2008; cf. Mitchell 2008). In this respect, Kötter's account of 'images' is obviously capable of conveying a truism of the discussion (cf. 4.1). Moreover, Kötter's account also succeeds in reducing the extension of the term 'image', and therefore the degree of complexity that casuistry is faced with, because it excludes the range of so-called "mental images" (Mitchell 1984, p. 507), which are certainly hard to attach to the idea of a manufactured "surface" (Kötter 2007, p. 164).

However, despite these positive results, there are also, as usual, disadvantages to be noted. One of the most striking ones is that the above list of features does not allow for any stance on the controversy that Hentschel (2000) has hinted at (cf. 2.1.2.3). In other words, Kötter (2007) seems not to take into account the idea that there might be different kinds of "representational system[s]" (Goodman 1976, p. 226). Ultimately one can easily apply his list of features to written words and sentences, as these are also inscribed on surfaces, visible to the unaided eye and created by human beings in order to represent something. So, at the end of the day, there would be further work to do, in order to match the offered conceptual constraints to the instances under discussion (cf. Kötter 2007).

Nevertheless, Kötter (2007) captures – though only in very general terms – prominent anchor points of the conceptual discussion. Accordingly, there are more refined analyses on each of the listed features, and often in some sort of combination: Some try, as I said, to dig deeper into point (4) and distinguish, for example, between "narrative and depictive [modes of] representation" (Walton 1976). Others lay stress on point (2) and (3) in the way that they analyse the conditions and effects of visual competence and cultivation (cf. e.g. Schier 1986). Others investigate into point (1), for example, in the way that they analyse the design of *particular* images and also the *method* of analysing the design of images itself (cf. e.g. Panofsky 1932ff.).

Most recently, we also find authors that have begun to write the history of the conceptual preoccupation with images – thereby adding a context and expiry date to each of the previously mentioned approaches. J. Steinbrenner (2009), for example, lines up what he refers to as 'classics' of the 'analytical tradition' – starting with E. Gombrich (1977ff.). A. Kapust (2009) proceeds analogously with what she identifies as 'phenomenological positions' – starting with E. Husserl (2006).

By contrast, J. Elkins (2011) refrains completely from mapping the field, claiming that "it is not possible to list concepts of the image" (Elkins 2011, p. 13). He offers three arguments in support of his stance (most of which are *ad hominem*, which is why it is actually difficult to react on them): First of all, he argues that the word 'image' has grown to be merely a vehicle of underdeveloped polarisations ("the visual is somehow outside of language", Elkins 2011, p. 15). In this connection, he also adds that historians especially are found to refrain from explicating their background conceptions ("Words such as <image>, <picture> and Bild [sic] work in art historical discourses as placeholders: [...], we don't expect them to carry much of the argument" Elkins 2011, p. 15). Finally, Elkins claims that the variance of concepts and the proclaimed ubiquity of its instances are both *symptom* and *trigger* of the so-called 'visual studies' as research and study programs, and therefore preserved, instead of resolved, as a source of motivation and institutional justification (cf. Elkins 2011, p. 15).

2.1.3 Lessons learned

Obviously, I would not go as far as Elkins (2011) (see above), since I claim to have shown that it is at least possible to list *strategies of conceptualisation*. But Elkins is certainly right insofar as one could add another shelf of literature to each of the points, and one could also try other or more detailed patterns for sorting the conceptual accounts of 'images'.

But whatever sorting system one may try, the challenge that this paragraph is devoted to remains: how to find a meaningful way of using the key terms against this fuzzy background. In response to this challenge, I claim, first of all, that there are two lessons to be learned from the preceding overview.

1. Research on 'images' is not necessarily on 'images' – because, apparently, this term is used to denote partially incongruent sets of instances. The same is to be expected from other technical terms (as claimed, e.g., by Gadamer 1972, p. 392).
2. The idea of finding a completely satisfying way of using the term 'image' thus seems to be doomed to end up either with some sort of 'terminological normativism' or with completely unspecific (therefore useless) constraints.

Then again, I also claim that the drawbacks of this situation (indicated in the second point) can be avoided by focusing on the virtues of the afore-described strategies (cf. 2.1.2.1ff.). These virtues comprise *tentative* (meaning *reversible*), *explicit* (meaning *debatable*) and *concept-focused* (meaning *specific* or *specifying*) accounts that guide the use of terms in a way that is open for course corrections. The 'bottom-up strategy', as I call this triad of virtues, would thus be designed to keep the use of conceptual tools (i.e. technical terms and higher systematic units, like compounds of definitions) both efficient and controversial at the same time.[2] By contrast, the 'top-down strategy' would be designed to stop the queries, either by 'naturalising' a hierarchy of usages (cf. 2.1.2.1), or by rejecting the explication of the associated conceptual frameworks (cf. 2.1.2.2). With this, however, top-down strategies have proven vulnerable to methodological objections (cf. 2.1.1; 2.1.2.1; 2.1.2.2). And so, within this thesis (resp. within the broadest test run framework here chosen), it makes sense to opt for the 'bottom-up strategy' – whose sole drawback is an air of incompleteness as a corollary of the trial-bounded perspective – and to work on its transposition into concrete operations. Following the profile devised, a step in this direction is now the introduction of further points of departure (conceptual as well as methodological ones) that have been elected for a test run in this thesis.

2.2 An epistemic account of picture-like means

The previous section concluded with a call for tentative, explicit as well as concept-focused frameworks, while the term 'framework' has been (and will further be) applied to each systematic unit that carries the line of reasoning in this work – from the technical terms and related strategies of use to the definition of problems and supporting apporaches. The accompanying message was that, as long as we do not aspire to make the so-called "visual studies" (Elkins 2003) or "*Bildwissenschaften*" (Sachs-Hombach 2003) the new cosmology, it would be appropriate to narrow down the matter to some extent, by now continuing with the central problem definitions (cf. 2.2.1). In order to come up with a more tangible account of the latter (ultimately offered in 2.2.1.3), a specification of 'epistemic' problems (cf. 2.2.1.1) and applicable approaches (cf. 2.2.1.2) is required.

---

2 Some might object that this simply shifts the problemof diversity to other levels, namely to the level of supporting arguments or meta-strategies. This is true insofar as the problem of diversity is to a certain extent iterative. However, it stands to reason that a heuristic specification of frameworks aims at nothing more than that: a *test run* of frameworks, which is itself only a test run, etc. In other words, I do *not* suggest ruling out the diversity, but treating it in a certain way, namely to analyse it in terms of test run cases and not in terms of puzzle pieces forwhich onewould assume only *one* appropriate background systematic (e.g. in terms of an 'original' framework; cf. 2.1.2.1).

2.2.1 Epistemic problems

There is – and this should no longer be a surprise – more than one idea of what is to be referred to as an 'epistemic problem'. S. K. Langer (1957) is thus right in saying:

> "The 'technique', or treatment, of a problem begins with its first expression as a question. The way the question is asked limits and disposes the way in which any answer to it – right or wrong – may be given" (Langer 1979, p. 3).

On this assumption, I will now flesh out the main problem descriptions, beginning with some general remarks on the attributive usages of the term 'epistemic' (cf. 2.2.1.1).

2.2.1.1 Defining 'epistemic'

The term 'epistemic' is commonly traced back to the Greek noun *epistēme* (ἐπιστήμη), from which historians assume that it once covered the semantic field(s) of what is now identified as 'knowledge', 'insight' or 'science' (cf. Hager 1971, p. 588–593). But, while the nominal form 'episteme' is hardly used anymore, and firmly historical in that sense, the attributive use of the derivative 'epistemic' has grown popular in the past years. Accordingly, we find various combinations of the 'epistemic-*xy*' type.

Some of these combinations are used to express whether certain kinds of entities deserve particular consideration from the viewpoint of a certain account of knowledge or knowledge acquisition. Such combinations are, for example:

> 'epistemic status' (Baird 2004, p. 40), 'epistemic relevance' (De Regt & Dieks 2005, p. 140), 'epistemic role' (Carusi 2008, p. 252), 'epistemic significance' (Perini 2012, p. 159), 'epistemic weight' (Baird 2004, p. 28), etc.

To reveal the 'epistemic-*xy*' of a certain kind of entity in that sense, one thus needs to find out whether this kind of entity is, or can possibly be, a systematic unit (resp. an "integral part", Perini 2005, p. 262) of the framework under consideration – in the same way as, for instance, *beliefs* are systematic units of analyses of *knowledge*.

By contrast, other combinations of the 'epistemic-*xy*' type are used to denote certain sets of entities that are considered to have (or have had) an impact on actual states of knowledge or processes of acquiring knowledge. Such combinations are, for example:

> 'epistemic objects' ("*epistemische Dinge*", Rheinberger 1992), 'epistemic artifacts' (Knuutila & Voutilainen 2002), 'epistemic styles' ("*epistemische Stile*", Boehm 2001, p. 43), 'epistemic warrants' (Carusi 2012, p. 112), etc.

Entities that are subsumed to one or more of the so-termed sets are thus identified as actual means and manners of acquiring knowledge – in the same way as, for instance, an atlas can be called a thing that extends my knowledge of several regions on this planet. Accordingly, the use of the term 'epistemic' in that second sense (esp. when it is attributed to nouns in the plural form) is to a certain degree justified empirically, that is with respect to the *actual* modification of a *certain* state of knowledge or research.

While there might be further variants circulating in the community, it is already possible to draw a connection between the differentiated usages of the term 'epistemic' and two different yet interrelated types of problem definitions: The first of the above-noted usages can be associated with *questions on the ingredients of conceptual frameworks*. One might ask, for example: "Can a visual way of acquiring a mathematical belief justify our believing in it?" (Giaquinto 2007, p. 1). Here the related problem to tackle would be the as yet unsettled relation between the notions 'visual way' and 'justification'. And if indeed the 'visual way' turns out to be compatible with or even integral to the targeted idea or methodology of 'justification', the 'visual way' will thus be considered as relevant, significant or weighty from this particular 'epistemic' perspective.

The second usage, on the other hand, can be linked with *questions on the ingredients of historical frameworks*. In this context, one might ask: "Which ideas lead to the order and use of radiographs today, how do radiographs foster the acquisition of knowledge, and how do they influence the treatment of sick people?"[3] (Hosten 2010, p. 141). Here, the problem is as yet an undocumented (his-)story about the subtexts of technological pervasion. And if it turns out that the studied objects (e.g. radiogaphs) have had an appreciated impact on the acquisition of knowledge, they may be valued as having been 'epistemic' items in the particular context under consideration.

As indicated before, the corresponding research interests neither rival, nor are they irrelevant to each other, since, on the one hand, a historical account may trigger the review of conceptual accounts so that one is prompted to ask whether the conceptual accounts need to be revised in the face of historical shifts and changes. On the other hand, a conceptual review can have an impact on how (hi-)stories are told, for example with respect to the chosen descriptors and classifiers. So, in fact, both scenarios will find further illustration in the literature review below (cf. 2.3.2; 2.3.3).

3 Translation of the following quote: „Unter welchen Vorstellungen werden heute Röntgenbilder in Auftrag gegeben und benutzt, zu welchem Erkenntnisgewinn führen diese Bilder, und wie beeinflussen sie die Behandlung von Kranken?“ (Hosten 2010, p. 141).

For now, the task is to turn these still rather abstract problem definitions into a feasible roadmap. For this purpose, the next thing to do is to clarify how the identified types of problems are dealt with in this thesis. This much I shall reveal in advance: Both types, the conceptual and the historical one, play a role in this thesis, since an integrated view seems to be a promising, perhaps even inescapable track to follow when it comes to shedding light on the state and status of visual devices in recent scientific practices and accounts thereof (cf. 2.2.1.2, 2.2.1.3, 4.1.4).

2.2.1.2 Analysing 'function' and functions

The previous section was devoted to the analysis of the different problem definitions and approaches that are found fused (and confused) under the common label 'epistemic'. The difference between conceptual and historical framework analyses was highlighted in this context along with the corresponding terminological specificities. In addition, I will now dig deeper into the common grounds of these approaches, following the idea that the shared interest in *functions* (here esp. of visual devices) may serve as a key to an integrated approach. But before I can expound the details of this idea, attention needs to be drawn to the notion of *function* and related confusions in the recent debate. This will help tailoring an approach to the epistemic problems assumed (cf. 2.2.1.3).

The notion of *function* has been tackled, among others, by R. Cummins (1975). About a decade after C. G. Hempel (1965) had published his view of *functional explanation*, denying it the status of a real explanation as compared to the DN model, Cummins picked up the line of discussion and presented his own version which has been widely recognised in the philosophy of science and especially in the philosophy of biology and the social sciences. Since *causal* explanation is rarely, if ever, an option in the latter fields, there was (and still is) substantial interest in the question whether *functional* explanation "is a genuinely distinctive style of explanation" (Cummins 1975, p. 757). And so, like Hempel before – yet along the lines of a counter statement – Cummins sought to find out whether the "functional language" (Cummins 1957, p. 746) and the corresponding "explanatory strategy" (Cummins 1957, p. 765), as used in the sciences, can be illuminated, and hence backed up, by virtue of an analytical approach.

At the beginning of his analysis, Cummins picks up a "function-ascribing statement" (Cummins 1975, p. 741) from Hempel (1965) for demonstration purposes:

> "The heartbeat in vertebrates has the function of circulating the blood through the organism" (Cummins 1975, p. 742).

Hempel's (1965) analysis of this statement is cited as follows:

> "The heartbeat in vertebrates has the effect of circulating the blood, and this ensures the satisfaction of certain conditions (supply of nutriment and removal of waste) which are necessary for the proper working of the organism" (Cummins 1975, p. 742).

With regard to Hempel's analysis and similar approaches found in the literature, Cummins argues that "almost every recent attempt to give an account of functional analysis and explanation" (Cummins, 1975, p. 741) can be condensed into the claim that "a function-ascribing statement explains the presence of the functionally characterized item *i* [here: the heartbeat] in a system *s* [here: the organism] because it has certain effects [here: blood circulation] on *s*" (Cummins 1975, p. 741). In other words, "philosophical research on this topic has almost without exception proceeded under the following assumptions:

> (A) The point of functional characterization in science is to explain the presence of the item (organ, mechanism, process or whatever) that is functionally characterized.
> (B) For something to perform its function is for it to have certain effects on a containing system, which effects contribute to the performance of some activity of, or the maintenance of some condition in, that containing system" (Cummins 1975, p. 741).

Now, what is problematic about this received account of functional explanation?

With regard to the example in question (i.e. the heartbeat in vertebrates), Cummins points out, first of all, that the proper working of the organism is *not necessarily* secured by the beat of a *heart*, since a *machine* can fully replace the heart. Therefore, he first objects that one cannot validly deduce (i.e. explain) the presence of a heart from certain effects (here: blood-circulation or working organism; cf. Cummins 1975, p. 743).

In the next step, Cummins shows that E. Nagel's (1961) defence of Hempel cannot compensate for this drawback. Cummins' already modified and customised version of Nagel's defence goes as follows:

> "Although it is true that the presence of a working heart is not a necessary condition of circulation in vertebrates under all circumstances, still, under normal circumstances – most circumstances in fact – a working heart is necessary for circulation" (Cummins 1975, p. 744).

This analysis fails as well, since, as Cummins claims, there are phenomena (described, e.g., in biology), where under *abnormal* circumstances there is still no surgery needed, because the required effects are "duplicated" (Cummins 1975, p. 745) by other instances. The kidneys, for example, and perhaps also some parts of the brain, can replace each other if one of them fails (cf. Cummins 1975, p. 745). Therefore, Cummins further objects that Nagel's account rests on a too rigid and thus ultimately false premise about fixed (resp. 'normal' or 'necessary') relations between effects and their causes. According to Cummins, this is only introduced as a premise for reasons of stipulation, that is, for rescuing the possibility of deducing the presence of an entity from its effects.

But these are not the only – though already compelling – objections to Hempel's analysis and Nagel's rescue attempt. On top of that, Cummins argues that there are always *more* effects than those that are chosen for a functional characterisation. Heartbeats, for example, produce a more or less regular sound scheme. Yet, the latter has been ruled out as an *irrelevant* effect in the example under discussion. This means, in turn, that a *choice* has crept into the scheme and calls for motivation:

> "What is needed to establish a selected-effects theory is a general formula that identifies the appropriate effects. [...] For no matter which effects of something you happen to name, there will be some activity of the containing system to which just those effects contribute [...]. Heart activity, for example, keeps the circulatory system from being entirely quiet, and the appendix keeps people vulnerable to appendicitis" (Cummins 1975, p. 752).

Cummins thus decides on discussing *evolution theory* as a possible candidate for the "selected-effects theory" – because evolution theory provides criteria, such as *fitness*, that seem to facilitate the matching of appropriate qualities, functions and contexts. However, since these qualities, functions and contexts need to be, as just said, *selected* as appropriate instances of a likewise *selected* criterion (here 'fitness'), the search for a higher selection strategy goes on from this point. It will certainly not be easier along the way, since there are, for example, "cases in which proper functioning is actually inimical to health and life: the functioning of sex organs results in the death of individuals of many species (e.g., certain salmon)" (Cummins 1975, p. 754f.). Hence, Cummins shows that the attempt to find a "selected-effects theory" eventually faces what might be called the 'highest rule dilemma' – that is, it ends up either with an *indistinct* or *incomprehensive* account of the relevant instances (here: effects).

In the end, Cummins (1975) thus comes to the conclusion that Hempel's (1959) and Nagel's (1961) analysis of functional explanation "fails to give any hint as to the point of identifying certain effects as functions" (Cummins 1975, p. 757).

Still, this does not signify the end of the entire project. Cummins suggests from the beginning that his criticism is not directed against the initial idea of an analytical approach to functional explanation. It shall rather serve as a reason to reflect on the flaws of and potential improvements on the former standard analysis.

Along these lines, Cummins comes up with the idea that the latter suffers from a confusion of the *explanandum* (i.e. the to-be-explained entity): It is not the *existence* of an entity that is meant to be elucidated by a "function-ascribing statement" (Cummins 1975, p. 741), but the very fact that this entity is capable of contributing to the accomplishment of a certain task. Cummins ergo proposes to analyse the *function*-ascribing statements in terms of *disposition*-ascribing statements (cf. Cummins 1975, p. 758).

In order to explain a *disposition* – that is, a capacity or potential of an entity in contrast to its sheer presence – Cummins considers it necessary "to explain how manifestations of the disposition are brought about given the requisite precipitating conditions" (Cummins 1975, p. 758). The precipitating conditions, in turn, are capable of being *precipitating* conditions with respect to their own precipitating conditions, etc., so that one obtains a nesting of ascriptions in which each functionally characterised component (including the components of those components, etc.) is earmarked as capable of contributing to the conditions that are required for the accomplishment of other conditions, etc.

This revised account might appear somewhat intricate and surely leaves a flank open as to the notoriously difficult concept of dispositions. Still, Cummins offers a valuable analogy for illustration purposes, which qualifies it at least as a palpable intuition:

> "Assembly-line production provides a transparent example of what I mean. Production is broken down into a number of different tasks. Each point on the line is responsible for a certain task, and it is the function of the workers/machines at that point to complete the task. If the line has the capacity to produce the product, it has it in virtue of the fact that when these tasks are performed in a certain organized way – according to a certain program – the finished product results. Here we can explain the line's capacity to produce the product [...] by appeal to certain capacities of the workers/machines and their organization into an assembly line" (Cummins 1975, p. 760).

Accordingly, by ascribing a function to an entity (e.g. to an assembly-line worker or to the heartbeat in vertebrates), this entity is marked as capable of fulfilling a certain task (e.g. tightening bolts or pumping blood). The respective task is defined by what Cummins calls the 'program' (e.g. the concept of assembly-line production or blood circulation), which defines the conditions of its own accomplishment (e.g. tightened bolts or circulated blood). The successful accomplishment of the program – that is, of the conditions or tasks that it defines – may at the same time have its own function against the background of another program (e.g. the assembly-line may contribute to the implementation of a production plan and the blood-circulation may carry parts of a vertebrate's metabolic system, etc.). And so, to cut it short: functional analysis is defined in terms of a matching of functional ascriptions here. The matching yields a functional explanation, if there is reason to believe that the functionally characterised entity is capable of accomplishing a particular task, because its own components are capable of securing the conditions for accomplishing this task, etc. The *empirical validity* of such functional statements or ascriptions, however, is none of Cummins' business, because, as he implicitly suggests, an analytical account is more resistant to the problem of ultimate justification when it offers a *model* of a valid functional explanation, instead of the highest rules of validity (on which Hempel and Nagel failed).

And, indeed, the major problems of the received view (i.e. the constant lack of higher justificatory instances, such as more general relevance-criteria, and the frequent struggle with empirical exceptions) seems to be circumvented here; because if one takes the *program* (e.g. the concept of an assembly-line or blood circulation) as one premise and the *confirmed accomplishment of its tasks* (e.g. the working assembly-line worker or the beating heart) as the other, then the function of the employed components (e.g. the worker or the heart) is validly explained within the framework so-devised. Moreover, due to the fact that *any* kind of program may fit the model, even sentences such as "The heart functions as a noise-maker" (Cummins 1975, p. 762) can make sense – for example, with respect to the functional program of a piece of music, where the heartbeat may contribute to the rhythm. Hence, apparently, Cummins' account leaves room for the diversity of empirical frameworks ('programs') and the simultaneity thereof.

Having thus captured some useful insights and virtues of Cummins' view of functions, it is time to mention that Cummins distinguishes another type of "functional speech" (Cummins 1975, p. 747), which is not covered by the afore-described account, as it is considered to be *categorically different*. In more precise terms: what has been said so far concerns the ascription of functions to "natural systems" (Cummins 1975, p. 747). This is considered different to the function of means in the context of "deliberate actions" (Cummins 1975, p. 747; also known as "pragmatic functions", Kötter 2007, p. 164, or "instrumental functions", Beyer & Lohoff 2005, p. 11).

The point about this distinction is not, as one might guess, that humans cannot be an instance of 'natural systems'. For if that were the case, then Cummins could not use the assembly-line as a paradigmatic example for the latter, since the assembly-line involves human workers as veritable functional components. Rather the point is that in the context of deliberate actions, we examine the *function* of an entity, "because we may wish to know *why* it was put there" (Cummins 1957, p. 746). Yet, by approaching that question – for instance, by pointing to the fact that the assembly-line was *installed in order to* produce cars – we neither specify the function of the employed item itself, nor the terms of its capacity, but we specify the *function of its use* (i.e. the function of an action) and the fact that *it occurs on purpose*, meaning that it is performed with respect to a certain *goal*, the achievement of which is the 'function' of this measure in that case. Thus construed, another explanatory strategy is indeed to be noted here, namely explanation with reference to *reasons*, which accordingly deserves a distinct category (e.g. "teleological explanation" Cummins 1975, p. 747).

It is important to keep in mind the difference between the two notions of 'function'. This difference is becoming ever more apparent in current research on visual devices where it is successfully blurred. Consider, for example, the following list of *functions* presented by A. Beyer and M. Lohoff (2005) as part of their collection:

> "[Visual devices] are found used for the purpose of objectification, furthermore for the purpose of specification, modification and (often temporary) documentation of knowledge, moreover for the recording and communication, structuring and explication [of information], [and they serve] as analytical tools, arguments, evidences and legitimacy."[4]

The categorical inconsistency of this list is plain to see against the background of Cummins' distinction: it makes a difference whether one claims that a visual device serves *the purpose of explication* (this claim needs to be justified with respect to the user's intention, which can be figured out by asking, *What do you need the device for?*), or whether one claims that a visual device is used *as an argument* (this claim needs to be justified with respect to the task that the device fulfils and that can be enquired by checking, *In what way can the device contribute to the accomplishment of a given purpose?*). From this angle, it seems thus safe to say that the term 'function' likely refers to *both* of the two different, afore-distinguished notions of 'function' in the list above.

However, one might want to avoid such a blending of notions as given in the list above for basically three reasons. First of all, technical terms (here: 'function') are commonly used to draw distinctions. This is what eventually makes them efficient

4 Translation of the following quote (information in square brackets added): „[Wissenschaftliche Bilder] dienen der Vergegenständlichung, der Wissenskonkretisierung und Formung, der (oft nur temporären) Wissensdokumentation, der Fixierung und Vermittlung, der Strukturierung und Klärung, als Analyseinstrument, Argument, Beweismittel und zur Legitimation" (Beyer & Lohoff 2005, p. 12).

or not, like in the above case where differences are blurred. Secondly, in the case under consideration, there would be enough fairly intuitive options for denoting the revealed differences (e.g. one could talk about 'functions' in the second and 'purposes' in the first case), so that the imprecision is hard to defend. Finally, and most importantly, the different notions of 'function' are linked to different methodological frameworks respectively, which run the risk of being confused in the same instance; because in order to find out about the *purpose(s)* of an entity, the corresponding user(s) needs to be consulted, or a testimony of the user's intention should be shown, whereas the *function* of the entity is clarified with regard to the task that the entity is capable of accomplishing from the viewpoint of a certain program and with regard to its components (see above).

Having thus gained a more differentiated view of what it could mean to approach the function of things, I will now be able to bring the threads together, in order to specify the agenda of this thesis. Accordingly, I will now further elaborate on the problem definitions (as distinguished in 2.2.1.1) against the background of the analysis of the concept(s) of function performed in the present paragraph.

2.2.1.3 Desiderata: epistemic weight and epistemic surpluses

In order to arrive at a more precise problem definition and solving agenda, I will now start over and try to tie the threads together at a more general level of abstraction.

I will begin with the supposed matters of agreement: there seems to be little doubt that the defining problem of the philosophy of science is finding an answer to the question, *What is science?* Starting from there, further questions are commonly defined (e.g. *How is it done correctly? What does it bring about?*) and rendered more specific, depending on the chosen frameworks and variables (e.g. depending on the instance that is focused on, or whether a descriptive or normative answer is targeted). At the end of the day, the scope of possible problem definitions is obviously broad enough to nourish an independent area of research and can therefore hardly be presented here in general terms.

At the most general level, one rather tends to group the answers to these questions, in order to characterise the area of philosophy by characterising the "*-isms*", "*views*" and other traits of argumentation that have come up. In this context, we can maintain (esp. in the first half of the 20th century; cf. 2.3.3) that many condensed the defining agenda of the philosophy of science into the problem of how sets of statements must be related to each other in order to acquire logically stable frameworks (e.g. in terms of explanations) and, in this sense, *scientific knowledge*, or that is to say, *scientifically acquired knowledge*. By starting at the level of statements, however, it was (and still often is) presumed that those are already given in a tractable form and that the controversy should start with respect to the types, standards and consequences of their relation to each other.

A somewhat *circular argument* for this presumption is that the genesis of statements is of no further relevance here, because all relevant systematic units (e.g. observations, models, laws, logical rules, etc.) can be represented by and as statements and because the genesis of these statements is itself not a relevant systematic unit. Such an argument can be found, for example, in H. Reichenbach's work (cf. Reichenbach 1938, p. 6ff.), to name but one of the early proponents (more information on his argument is provided in 2.3.2). In fact, the only or most efficient way of tackling an argument like this is by getting granular on the associated criteria of systematic *relevance*, since these likely involve a more substantial underpinning – for example, in terms of a debatable premise on the defining modules of scientific knowledge.

Such a *dogmatic argument* – which is at least questionable, after all – was, for instance, presented by H. Poser (2001), who starts his introduction into the philosophy of science ("*Wissenschaftstheorie: Eine philosophische Einführung*") with a part on "Language as a starting point" ("*Die Sprache als Ausgangspunkt*", Poser 2001, p. 28), where he states that science is given to us as a linguistic compound and, moreover, that thoughts and insights are only conceivable of via linguistic means:

> "If we choose language as the starting point of our endeavour, this gives us not only the advantage that we can work on the linguistic material that defines the sciences, but it also has a deeper epistemological reason: [...] we cannot get a grasp on thinking other than by linguistic means; even if there are lower forms of non-verbal thinking, it is not possible to conceptualise abstraction without involving language"[5] (Poser 2001, p. 28).

On such a basis, of course, one has the legitimisation to switch right to the next level of systematic units – in Poser's case, to the syntactic, semantic and pragmatic entities that compose propositions as well as their scientifically established relations (cf. Poser 2001, p. 29ff.). Recent authors, however, added a question mark

5 Translation of the following quote (square brackets added): „Wenn wir die Sprache als Ausgangspunkt unserer Untersuchung wählen, so hat dies nicht nur den Vorzug, dass Wissenschaften als sprachlich formulierte Gebilde vorliegen und wir deshalb an dieses Material anknüpfen können, sondern auch einen tiefer liegenden erkenntnistheoretischen Grund. [...] Nun kann man aber des Denkens gar nicht anders als sprachlich habhaft werden; und selbst wenn es niedrige Formen sprachlosen Denkens geben mag, ist Abstraktion ohne Sprache nicht vorstellbar" (Poser 2001, p. 28).

to this received point of departure in the philosophy of science (cf. 2.3.2; 2.3.3), because they assume (as I do) that the footing on which it is built, and which has been exemplified by a quote from Poser (2001), is ultimately false. To elaborate on this intuition, the scope of the problems needs to be shifted or extended. Hence, in addition to the analysis of the functional ingredients of scientific knowledge (e.g. in terms of observations, models, laws, etc.), analysing the functional ingredients of those ingredients has been suggested, also with special regard to *visual devices*. Since if it turns out that the latter can have an essential function in scientific contexts and that they are not linguistic entities in the presumed sense, then the received view must be rejected.

Based on these assumptions, it is now possible to specify the central questions of this thesis, namely (1) the question on the epistemic weight and (2) the question on the epistemic surpluses of visual – and especially *picture-like* or *pictorial* – means.

The question on the *epistemic weight* of visual devices shall be interpreted as the question whether visual devices can be functionally involved in procedures that in turn can be functionally involved in science (cf. 4.2). I will discuss this first question on the basis of L. Perini (2002ff.), who proposes an affirmative answer, claiming that visual means have the capacity to be integral to *arguments*, which in turn can be integral to other systematic units of the scientific acquisition of knowledge, such as *explanation* (cf. esp. 4.2.1)

The question on the *epistemic surpluses* shall be interpreted as the question whether visual devices have specific functions or unique potentials that other devices – especially 'linguistic' devices in a certain, opposed sense – could *not* have (had) and that thus extend (or have extended) the scope of purposes that is pursued in exemplary scientific contexts. This second question will be discussed with reference to various authors, including those who actually employ visual means (cf. 4.3).

The expected outcome is a plain answer to the first question, given a limited set of frameworks that specify the functional components of pictorial means and of arguments alike, so that these sets can be scrutinised with regard to conceptual overlaps. By contrast, an open list of answers is expected with regard to the second question.

Why an *open* list in the second case? Well, given that science cannot (yet or ever) be described as a functional program in the sense of Cummins (1975), that is, as a series of tasks that need to be fulfilled in order to guarantee its capacity to produce knowledge (cf. 2.2.1.2), we rely on historical research that sheds light on the conditions that may have paved the way to punctual, context-bounded notions about success or progress according to the testimonies that are given credence in recent contexts.

Argumentative support for this disclaimer comes from P. Feyerabend (2010), who argued that we cannot extrapolate from one 'scientific success' to others, *precisely because* we cannot conceive of science as a fixed and finite entanglement of functional modules:

> "The idea of a method that contains firm, unchanging, and absolutely binding principles for conducting the business of science meets considerable difficulty when confronted with the results of historical research. We find, then, that there is not a single rule, however plausible, and however firmly grounded in epistemology, that is not violated at some time or the other. [...] More specifically, one can show the following: given any rule, however, 'fundamental' and 'rational', there are always circumstances when it is advisable not only to ignore the rule, but to adopt its opposite. For example, there are circumstances when it is advisable to introduce, elaborate, and defend *ad hoc* hypotheses, or hypotheses which contradict well-established and generally accepted experimental results, or hypotheses whose content is smaller than the content of the existing and empirically adequate alternative, or self-inconsistent hypotheses, and so on" (Feyerabend 2010, p. 7).

Accordingly, the idea of an integrated – both conceptual and historical (cf. 2.2.1.1) – approach does not only make sense against this background, but it seems inescapable when it comes to shedding light on the epistemic weight and surpluses of picture-like means in the proposed sense; because in order to prevent the required performance and matching of functional analyses (in response to the question: *What are visual devices capable of contributing to?*) from becoming a boundless pen and paper exercise, eventually resulting in over-generalised or entirely insignificant accounts of scientific practices, historical frameworks must be consulted, in order to give investigative priority to those functional 'programs' (in the sense of Cummins 1975; cf. 2.1.2.2) that are employed on purpose, with all the limiting consequences that arise from such a restriction and delegated approval of reference contexts and (hi-)stories.[6]

---

6 The major consequences, which have coined the philosophy of science in the previous years, are that accounts of scientific practices are now bound to more descriptive perspectives and local scopes of validity (cf. 2.3.2). Even former standard assumptions (e.g. on the importance of *arguments* in the scientific processing), must face the suspicion that there are exceptions to it. P. Feyerabend (2010), for example, doubted the importance of arguments in science, claiming that views and contentions are not in any case incited and driven by arguments (cf. Feyerabend 2010, p. 8f.). Consequently, the focus on the role of visual means in the context of arguments relies itself only on

In face of these constraints, the following paragraph will introduce conceptual aids for a first heuristic account – an identikit picture, if you will – of the exemplary contexts (resp. the associated [hi-]stories) that are worth an 'autopsy' (cf. Boehm 2007, p. 41) in the light of the afore-distinguished questions.

2.2.2 Figures and picture-likeness

G. Boehm is right in one respect and wrong in another, when he claims that the modern understanding of visual means (for him "iconic figurations"; "*ikonische Figurationen*", Boehm 2007, p. 37) draws not only strongly on linguistic and rhetoric concepts, but also on the results of research on perception (cf. Boehm 2007, p. 37). Regarding what is wrong or at least controversial: I must say that I do not see much support for the assertion that the different views of visual devices accumulate into a 'modern understanding'. This seems to be a projected result rather than the starting point of the recent debates. On the other hand, Boehm is certainly right in that many authors adjust their views – not the least due to the *lack* of a cohesive understanding – to various fields of research. For this reason, it makes good sense to revisit the concepts that guide the casuistry – especially the target identification and profiling – in those fields that are committed to "visual [framing] devices" (Idhe 2009, p. 205), "optical devices" (Hunter 2010, p. 193) or "visual practices" (Elkins 2010).

The analysis of the field is complicated by the usual terminological peculiarities and mannerisms (cf. 2.1), which offer a distinct type of problem, not just a notational collision or mismatch in the hermeneutic realms. For instance, one finds that many authors employ the terms 'visual' and 'optical' – which becomes an issue, if my cup of coffee eventually counts as a 'visual' device as well. On the other hand, not as many of the authors adopt all the sub-categories that have been introduced into the debate – like "visual representation" (Robin 1992, p. 11), "pictorial representation" (Schier 1986) or the above-noted "iconic figurations" ("*ikonische Figurationen*" Boehm 2007, p. 37), not to mention the reputed *sub-sub*-categories, like "diagrams, maps and models" (Goodman 1976, p. 170); this becomes an issue when it comes to discussing arguments that are devoted to different scales.

---

a locally justified view and test-run case (as announced before; cf. 2.1.3), wherefore I emphasised the condition of individual or local identification with the agenda proposed in this thesis (cf. preface).

The task of the coming sections will therefore be to narrow down the conceptual apparatus – that is, the set of descriptors, classifiers and background information – that has been chosen to guide the selection of examples and that will help render them debatable in terms (resp. in *the* terms) of the initial questions and dedicated approaches.

2.2.2.1 What are scientific figures?

Figures are known in or from almost all fields of the liberal arts in the broadest sense, including dance, sports, games, music, sculpture, painting, literature, geometry, logic, astronomy, and more. Correspondingly, there is some room for bending the idea of what a 'figure' is. Some, for example, apply the term 'figure' to *material* or *plastic* items, like statues, game pieces, etc. Others blend it with the scope of the term '*form*' – thus capturing also syllogisms, rhetorical devices, musical schemes, etc. Moreover, some think of figures as being, *solid, fixed* or *static* entities, such as sculptures, constellations, geometrical shapes, etc., while others also adopt rather *variable* or *ephemeral* entities into the family, such as shapes in clouds or sand, etc. Finally, to put a preliminary end to this list, figures are also found incorporated in scientific publications and other documents that originate in academia – while the conditions of incorporation (esp. the functional profiling and methodological embedding of figures *as* scientific and possibly picture-like figures) are subject to discussion here (cf. 2.2.1).

From a taxonomical point of view to begin with, one might be interested in the question *What do figures that are incorporated in scientific documents have in common with those figures that occur (or wander) elsewhere?* In this regard, the history of the term 'figure' and related concepts – which has been presented in great detail by E. Auerbach (1967) and brought into the discussion, for instance, by G. Boehm (2007) – provides several conceptual recurrences which can be abstracted and condensed into an *idea.*[7]

For example, it has been suggested that a figure usually appears as a composed unit ("*ganzes Gefüge*", Boehm 2007, p. 38) which obtains a distinct shape either by standing out (active definition), or by getting discriminated (passive definition)

---

7 Of course, the more an account seeks to be comprehensive, the more it runs the risk of being anachronistic, normative or superficial (cf. 2.1.2.1). It is thus important to take the definition not as a *concept explication* in the strict *exegetic* sense, but as a *framework genesis*, which is inspired by various sources, but eventually validated with respect to only some exemplary ones (cf. 3.1f.).

against a given background (cf. Boehm 2007, p. 35ff.). Particular emphasis is thus laid on the relation between a figure and its context, which is specified as a "contrast" (Boehm 2007, p. 35ff.), and as such regarded as ontologically relevant in the sense that a figure and its background are constituted in the light (resp. shade) of each other, whilst remaining essentially different (cf. Boehm 2007, p. 35ff.).

To put it in the terms of Boehm (2007):

> "The *figura* stands out from its context, for example a speech. This is the only way it receives a unique and distinct profile. It must appear different enough, in order to be identified as a figure"[8] (Boehm 2007, p. 35).

With regard to the consulted scientific documents (cf. 3.1f.), one could add as support that those figures are profiled, too, particularly for and via being *inscribed* and *indexed* (cf. III, intro.), which fits well with the intuition of them being distinct units.

Yet there is more about the relation between figure and background. Some fuse the ontological view with a *technical* prospect, which means that they expect the relation between figure and background to be to a certain extent disposable for further arrangements (static version) or interplays (motional version) and for exploitations of the whole unit. G. Brandstetter (2007), for instance, argues with respect to figures in ballet dance:

> "The arabesque which is a distinct figure in ballet dance cannot be fully accomplished or employed as a scheme. The figure defies such an access – the dancer can only traverse the figure and, thereby, [s/he] de- and transfigurates it. Seen this way, 'figuration' denotes the scope of possible realisations of a figure [...], which a dancer can merely transit without actualising all of them"[9] (Brandstetter 2007, p. 13f.).

Hence, Brandstetter combines an ontological view with a technical modularisation of the figure-background relation when she describes figures as constituted – that is, actualised (whether only approximately or not) – within the flow of a dance performance.

---

8 Translation of the following quote: „Die ‚figura' hebt sich vom Kontext, z.B. der Rede, ab. Nur so gewinnt sie Eigenart und ein deutliches Profil. Sie muß *anders genug* erscheinen, um als sie selbst erkannt zu werden" (Boehm 2007, p. 35).

9 Translation of the following quote: „Die Arabesque, als eine der herausgehobenen Figuren des Balletts, ist nicht ein erfüllbares oder ausfüllbares Schema. Vielmehr entzieht sie sich als Figur – der Tänzer kann sie nur durchqueren und damit immer schon de- und transfigurieren. So betrachtet bezeichnet „Figuration" den Möglichkeitsraum einer Figur [...], durch die der Tänzer hindurchgeht, ohne sie vollständig zu besetzen" (Brandstetter 2007, p. 13f.).

Thus construed, Brandstetter's account also bears comparison with others' accounts, if you consider (for the price of an anachronism), Lukrez' notion of atoms as uniting and de-uniting 'figures' ("*figura*", Auerbach 1967, p. 59), which compose variable kinds of distinct compounds (i.e. smaller or larger atoms) in the course of their constant drift and interplay, which E. Auerbach refers to as a 'dance' as well ("*Reigen von Figuren*", Auerbach 1967, p. 59), and which are – as disposable particles – affined to the technological paradigm from the 20th century on at the latest. So, the point is that the technogenic view of the figure-background relation implies that these compounds are conceived of as *means* or *products*, even in Brandstetter's example of ballet, where figures tend to defy, yet still vaguely allow for a technical approach. This seems to capture another stable intuition, apart from them being distinct units.

Now, when looking at this collection of ascriptions and attributions, it seems to me that it is, indeed, possible to arrive at an inclusive concept-based account of figures without bending the cited approaches too much at the relevant hinges. In other words: as part of the first round of framework specification (further adjustments will follow in 3.3 and 4.2f.), I will suggest the following notion of figures which is at least inspired by, if not coherent with, the accounts of Boehm (2007) and Brandstetter (2007): A figure shall be defined as the *whole* of a profile, contour or shape, which is referred to as a distinct entity or which can be ascribed to an entity as one of its distinct features. The distinctness of the figure depends essentially on its context so that it can be experienced as switching, transforming or vanishing in case the context changes, for instance in case the context varies in terms of colour, sound, shape, motion, lighting, perspective, cognitive dispositions, background knowledge, modes of exploitation, and more. In science, this 'background' is apparently given in terms of inscriptions, indexes and strategies of framing and separation from the text body, to name merely the graphically dimensioned measures of constituting figures (cf. III intro). A 'scientific figure' is thus a functional unit, which is established against the background of certain (to-be-explored) treatments.

Thus construed, a figure is by no means a natural kind (cf. Bird & Tobin 2008) in the sense that it would denote a context-invariant entity; even more so if one counts in a *user*, who constitutes the relation between a figure and its background.[10]

---

10 This condition is essential in order to capture the intuition that figures may be part of psychological tests, children's games and training programs, where the participants need to prove that they can apply certain education-dependent classifiers – which requires more than just a sensory apparatus.

This assigns figures to the group of entities that are *approached* as instances of a certain type of reference system (e.g. conventions) rather than merely detected as a certain type of sensory input.

Now, even if this definition covers major parts of the semantic spectrum, a meaningful investigation and discussion "*ad exemplum*" (Boehm 2007, p. 41) requires *further specifications*. The following remarks (1) and (2) shall provide the corresponding rationale and thus bridge between this and the next section:

(1) The scope of examples, as discussed below (cf. III), is more specific compared to the definition of figure suggested above. To avoid the impression of arbitrariness, further distinguishing marks thus need to be added – especially also with regard to the controversial conditions of *picture-likeness*, which probably apply to only some of the figures under study. One could think of achieving this very efficiently by simply drawing focus on all items that are referred to as 'picture-like figures' in the cases under study. However, such a *case-based specification* implies giving up the virtues of conceptual explicitness (cf. 2.1.2.2); but they are needed in order to compensate for the next problem (noted in point [2]).

(2) The term 'figure' is only rarely found used in the targeted controversy. Therefore, in order to create some anchor points in terms of shared vocabulary, other terms need to be introduced, either as synonyms or – and that would fit the first problem – as vehicles of conceptual add-ons. Promising candidates in this context are the terms 'picture' and its derivate 'picture-likeness' since, on the one hand, these terms are common in the reviewed accounts (cf. esp. Perini 2002ff.; cf. 4.2.2). On the other hand, the conceptual correspondences between figures and pictures will allow us to draw and to maintain the links to the more general aesthetic debates, which offer access to long-established reflections on pictures. On that note, I would like to continue upon the following rationale:

> "Figuration is not sufficiently specified as something separable or schematic. [...] Only if one realises the entanglement between [the process of] figuration and the structure of a *picture*, one gets a proper grasp of its status"[11] (Boehm 2007, p. 35).

---

11 Translation of the following quote (square brackets added): „Figuration wäre nicht zureichend beschrieben, als etwas Isolierbares oder Formelhaftes [...]. Erst wenn erkannt wird, dass sich das Figurieren mit der Struktur des Bildes selbst verschränkt, kommt sein wirklicher Status zur Geltung“ (Boehm 2007, p. 35).

2.2.2.2 What makes scientific figures 'picture-like'?

Figures and *pictures* are not treated as identical in the literature under consideration (cf. Boehm 2007, p. 36), even though they share some class-defining features, such as *being perceptually accessible, being practically established* and *being integral to human modes and systems of expression.* However, despite these common elements, the distinction between figures and pictures can be defended, because, to offer but one difference, figures are found *indexed* – not only but also in the scientific realms – while the *pictures* that result from the inspection of figures and the figurative scenarios that may be drawn from the study of pictures both seem to refuse this treatment, that is, they are not as easily encircled and listed. The question now is how to substantiate this intuition without anticipating too much of the analysis below (cf. IV).

If we recall the notion of *figure* as outlined before (cf. 2.2.2.1), emphasis was laid on the claim that figures are *composed units* ("*ganze[...] Gefüge*", Boehm 2007, p. 38). From this we may derive that in case the units can be broken down into *further* units – until they occupy whole sequences of apperception and examination – they move away from being *merely* figures towards being more complex compounds (e.g. *pictures*, *dances*, *speeches*, etc.). Accordingly, one way of fleshing out the distinction between figures and pictures is to set a difference in complexity and hierarchy, such that figures are possible components of pictures, perhaps analogous to the relation between game pieces (figures) and whole games (pictures). However, since this analogy would also fit the relation between letters and texts, notes and symphonies, etc., a more specific idea of 'picture-likeness' has still to be achieved.

Yet this is easier said than done. There are many reasons why it is difficult to define 'picture-likeness'. One (rather strategic) problem is that this concept is discussed below (as part of Perini's framework; cf. esp. 4.2.2), which is why it makes sense to refrain from early pre-configurations. Another problem in this context is that this concept is of major systematic importance here, for it may establish a bridge between historical accounts of *figure*-based practices and epistemological accounts of *pictures*; because only if these domains are brought in touch with each other, will the sense of the current buzz around the epistemic status of 'scientific visualisation' open up. Indeed *as* picture-like items, figures help to prove language-biased views of science wrong (cf. 2.2.1.3, 2.3, IV).

When weighing the drawbacks of a pre-definition – that is, the overly reduced complexity and the dogmatic overtone – against its systematic relevance, it seems to be convenient as a compromise to anticipate some of the *affirmed* elements of the definitions debated, for these elements allow us to identify the chosen examples (resp. their descriptions; cf. III) as veritable targets of the later analysis without making this analysis obsolete. Moreover, a pre-definition paves the way for the introduction of motivators, as those are expected to back up the research agenda also with reference to conceptual issues (cf. 2.3).

Thus, let us consider the following heuristic specification as a part of the test-run framework, which is further explored in this thesis: an entity is *prima facie* considered as 'picture-like', if it is approached as an instance of a topological, geometrical or graphical coding system. For example, one could read out a text without mentioning where it is located on a book page and how small the letters are, but one might want to delineate the architecture of a diagram, in order to grasp and distribute the information that it conveys. Hence, the attribute 'picture-like' specifies a certain type of *spatially attentive approach*, rather than the substance of the so-approached entities, which also implies that it is not possible to compile an inventory of picture-like items without getting into their origins and treatments (cf. III, introduction).

Now, as far as I can see, this pre-definition is prone to two objections. First of all, one could object that a *spatial definiens* (which is reportedly favoured in the cognitive sciences; cf. Maynard 2009, p. 83) is nothing new; even G. E. Lessing (1964) referred to the category of *space*, when he distinguished between the visual arts and poetry. This first objection which blames the lack of innovation is usually valid, though usually irrelevant; even more since the purpose of this thesis is to search *all over* for the best strategy to answer the initial questions. The second objection is far more interesting: one could object that the definition is blind to the spatial dimension of non-pictorial (e.g. linguistic) systems (cf. Krämer & Totzke 2012). I fully agree on that. Then again, I claim that this objection is weakened when taking 'picture-like' as denoting a certain type of *approach* (i.e. the choice of a certain coding scheme; cf. 4.2.2.4) without excluding the facts that there are 'hybrid' approaches (e.g. in case of comics, where the size of letters can be part of the coding system) and that one can try different approaches towards graphical instances. Thus construed, the category would also help to illuminate the *incentives* of the debate, which imply that figures are in part approached *as if they were pictures*, but *still* integral to science, giving rise to scepticism towards the received views of purely language-based science.

2.3 Picture-like means in science as a philosophical problem

This paragraph will add three motivators for analysing the use of picture-like means in science: The first motivator (cf. 2.3.1) justifies the distinction between language-like and picture-like means of representation by revealing the drawbacks of a purely linguistic account of representation. The second motivator (cf. 2.3.2) supports a science- and case-related analysis of picture-like means with regard to current discussions in the philosophy of science. The third motivator (cf. 2.3.3), finally, launches a history-focused argument, claiming that the design and use of picture-like means has grown into a techno-scientific format and, thus, into a subject of interest in the philosophy of science.

2.3.1 The failure of the purely linguistic view of representation

The first motivator is drawn from two sources: G. Boehm (1994) and W. Mitchell (1994) who detected a growing interest in picture-like means and suggested that it has to do with the limits of *language-focused* (resp. *picture-blind*) approaches both in philosophy and in the cultural sciences (cf. Boehm 1994; Mitchell 1994). But before I subscribe to their version of contextualising and promoting the observed shift or extension of interest, it is worth profiling at least *one* of the outdated accounts in order to reveal the conceptual quandaries against which the turn to pictorial means (e.g. as tools of scientific relevance) can, indeed, be seen as a valuable move.

Among the language-biased accounts mentioned by Boehm (1994) and Mitchell (1994) there is what I call the *linguistic view of representation.* This category is meant to capture all accounts of representation that bear on *linguistic* examples, premising that linguistic examples are the best examples when it comes to analysing the notion of representation. This preference for linguistic examples may even go so far as to say that *only* these types can exemplify what the term 'representation' should refer to. And this is exactly what applies to the *purely* linguistic view of representation, which correlates with this belief: *if we know how texts and talks represent, we also know if and (if so) how paintings, pieces of music, dances, masks, etc. represent.* What thus distinguishes the purely linguistic view of representation from other views of representation is the *ab initio* approval of a certain generalization, namely of the premise that language is *the* paradigm that helps to elucidate all that there is to know about representation.

Before coming to a 'real life' example of the purely linguistic view of representation, it will be helpful to first reveal its implications for this study. As far as I can see, three implications are of further interest, while only the first one is an implication in the strict sense; the other two may occur where the first one is defended:

(1) 'Representation' is a linguistic category. Hence, it applies solely and principally to linguistic entities or features thereof.
(2) The differences between the entities involved in a representation are of no further relevance to the representation, as long as they keep condition (1) satisfied.
(3) The features that profile a representation according to (1) also apply to entities that are not considered as linguistic entities in other conceptual frameworks.

Implication (1) allows the concept of representation to be treated as equivalent to a linguistic concept, for instance to the concept of denotation. Implication (2) excludes any specificity that could justify a sub-distinction between the entities that exemplify a representation, for example between language- and picture-like means of representation. And implication (3) opens the door for stipulated definitions, such as for the claim that so-called 'pictorial means' are also language-like in a certain (relevant) sense.

Now, the purely linguistic view is worth considering in this case because it has the potential to render the specific focus on pictorial means of representation obsolete. On the one hand, it may be that pictorial means are not subject to the problems under study, as they do not fit the conditions of representation (cf. [1]). On the other hand, there may be no significant difference between linguistic and pictorial entities with respect to their potential to establish a representation (cf. [2], [3]). In view of these concerns, I find it necessary, firstly, to claim that pictorial entities can be representations and, secondly, to defend the view that there is a distinction between linguistic and pictorial means.

The methodological motivation for this approach can be articulated in one sentence: We expect that a technical term such as 'representation' does indeed draw a distinction, and that it draws a distinction in a way that conveys common intuitions. In other words, 'everything-is-like-x' approaches are hardly ever illuminating, especially when they blur distinctions that are drawn in ordinary language. In fact, 'everything-is-like-x' approaches can even be misleading as in the case of the 'everything-is-like-language' view of representation, whose main advocate, Wittgenstein (in his early work, introduced below), turned casually blind to the fact that he *relied* on distinct representational capacities of *non-linguistic* means. This reliance will be fleshed out in the following section, in order to provide the first motivator for further research on pictorial means in general, as well as with particular respect to science (cf. esp. 2.3.2). The work of Wittgenstein offers valuable points of reference in this respect, since he tried to defend what I qualify as a purely linguistic view of representation even against his *own* intuitions, so that the rise as well as the fall of this view can be retraced using his work as an example.

The *rise* of the purely linguistic view of representation (at least of one rather explicit form of it) is historically associated with the *early* notes of Wittgenstein, particularly with the *Tractatus logico-philosophicus* (TLP) (cf. Dummett 1993, p. 27). Wittgenstein therein aims to clarify the relation between linguistic facts (propositions) and facts in general. More precisely, he aims to clarify whether linguistic facts can adequately and unmistakably represent other facts and, if so, under which conditions (cf. TLP preface). The clarification is supposedly driven by the questions of whether and to which extent we can *say* something about the world, given that we neither want to say something wrong, nor misunderstand each other. Thus, it appears that Wittgenstein tackles an old problem in the latest guise, so to speak, namely the relation between subject and world, which he specifies (in the first shot) as a relation between propositions and other facts.

In his first response to these problems, Wittgenstein appears to contend that linguistic facts *can* represent other facts provided that both correspond formally to each other (cf. TLP 2.13ff.). This condition of representation, which I identify as 'formal correspondence'[12], is quite controversial, as it can be broken down into further conditions. It appears to me that it is meant literally: a linguistic fact corresponds formally to another fact, if it has certain properties in common with the other fact (cf. TLP 2.13ff.). What might seem a harmless construal already rests on three conditions, namely

(1) linguistic *facts* represent linguistic or non-linguistic *facts*, if
(2) they have certain properties in common, and if
(3) they are distinct entities, meaning that a fact does not represent *itself* (cf. TLP 3.332).

These conditions form the 'skeleton' of Wittgenstein's account of representation.

For an in-depth analysis, one needs to specify what is meant with 'having certain properties in common'. The standard references in this context are E. Stenius (1969) and W. Stegmüller (1969), who offered an interpretation that was helpful to many exegetes. Stenius and Stegmüller famously argued that Wittgenstein's notion of representation comes closer to *isomorphism* than to being-a-proxy-for-something-

12 Wittgenstein uses the term "depiction" („*Abbildung*") (TLP 2.2ff.), but I think 'formal correspondence' captures the condition of representation quite well, and it makes the notation more comprehensible for an unexercised ear, especially since I follow Stenius (1969) and Stegmüller (1969).

else (cf. Stenius 1969, p. 121ff; cf. Stegmüller 1969, p. 539ff.). Strictly speaking, isomorphism is this type of relationship that holds between two non-empty ordered sets of elements *A* and *B*, just if *A* is a one-to-one (bijective) map of *B* (cf. Steiner 1976, p. 627–628; cf. Frigg 2011). *A* is a bijective map of *B*, if and only if the components of *A* relate to each other just as in *B*, so that *A* can be mapped onto *B*, while the structure of both is preserved. And more precisely, A is a bijective map of B (i.e. A is'isomorphic' to B), just if the components of A satisfy the relation R_A and the corresponding components of B satisfy R, where R is the relation in B corresponding to R_A (cf. Frigg 2011).

Buying into this analogy, Wittgenstein's idea of representation implies that linguistic facts are composites and, therefore, have the capacity to *map* – that is to say 'represent' – the relations between the composites that compose the world.

Having thus pinpointed the notion of representation in its broadest sense, it seems to be time for an example. However, it is *impossible* to provide an example for a representation, since, if we buy into the notes of Wittgenstein, a representation cannot be represented by itself (cf. TLP 3.332). And, indeed, one could state over and over again that a fact *A* represents a fact *B* in accordance with the required conditions, but these statements are not examples of representations, but examples of *statements about* representations. Yet, how can one provide a demonstration of a representation, if not like this? The key lies supposedly in the concept of *demonstration*. The following notes of Wittgenstein fit into this path: "A proposition [i.e. a linguistic fact] *shows* how things stand if it is true. And it *says* that they do so stand" (TLP 4.022).[13] On that note, the best exemplification of a representation appears to be an exemplary description of how one can show or, at least, of how it is shown, whether a statement represents a fact, that is, whether a statement is *true* (or not), since only a statement that represents is true (cf. TLP 2.21).

Now, this seems easier to accomplish. For example:[14] the statement 'Plato loves Socrates' represents a fact, just if the stated fact, namely that 'Plato loves Socrates', maps an actual composition of those things – in this case persons, emotions, etc. – that one refers to as 'Plato', 'love' and 'Socrates'. One can actually check whether the statement corresponds to what is the case, for example, by asking Plato about

13 Translation by D. Pears and B. McGuinness of the following quote: „Der Satz *zeigt* seinen Sinn. Der Satz *zeigt*, wie es sich verhält, *wenn* er wahr ist. Und er *sagt, daß* es sich so verhält" (TLP 4.022).

14 This example draws from B. Russell's introduction to the 1st English edition (TLP 1922).

his feelings towards Socrates. Depending on the answer, one will be able to infer whether Socrates and Plato are related to each other in love. In case one cannot infer their relationship, one needs to restate the question, or accept that the proposition cannot be verified[15] under the present conditions. To cut a long story short: linguistic facts (propositions) are established via verbal expression, but their 'representing-other-facts' (whether propositions or not) is not established in this way, but revealed (within the empirical boundaries) via inference and understanding (cf. TLP 4.02ff.). Still, Wittgenstein rests on *linguistic* examples, since a "gramophone record, the musical idea, the written notes, and the sound-waves, all stand to one another in the same internal relation of depicting that holds between language and the world" (cf. TLP 4.014).[16]

Before I move on to the problem that arises when maintaining the priority and exclusivity of linguistic examples, I will sum up the ingredients of Wittgenstein's notion of representation, because in this context it becomes clearer why this notion of representation is (not always, but often) deemed to be a *purely linguistic view*.[17] Here is a brief review:

(1) The term 'fact' denotes composites of things of whatever kind.
(2) The term 'representation' denotes isomorphism between two or more facts.
(3) Linguistic facts can be an isomorphic map of linguistic or non-linguistic facts.

Independent of whether isomorphism is the suitable category here, the *third* premise is the one that makes Wittgenstein's view of representation a *linguistic view* in the first place; since the third premise suggests – to put it more bluntly – that the world can be represented via *saying* something about it. 'Saying something about it' means to make statements, and statements are facts (i.e. composites of linguistic

15 E. Stenius (1969) hints at the problem that the terms 'true' and 'false' cannot be incorporated into a proposition that is isomorphic to another fact, as it makes no sense to represent 'false facts' (p. 128f.). Yet, as most of the exegetic issues, I must leave this open for a later discussion, as it distracts from motivating further research on pictorial means, which is the sole intention of this paragraph.

16 Translation by D. Pears and B. McGuinness of the following quote: „Die Grammophonplatte, der musikalische Gedanke, die Notenschrift, die Schallwellen, stehen alle in jener abbildenden internen Beziehung zueinander, die zwischen Sprache und Welt besteht" (TLP 4.014).

17 J. Steinbrenner (2009), for instance, doubts that Wittgenstein considers pictorial means as entirely language-like (cf. Steinbrenner 2009, p. 286). A. Danto (1994) and D. Mersch (2006), on the other hand, affirm that Wittgenstein offers a purely linguistic view.

entities) that can map other facts (i.e. composites of linguistic or non-linguistic entities). The mapping is possible, since all of them are facts. They share their 'being-a-composite' and, for this reason, they can be compared in regard to 'being-composed-in-the-same-way'. But, if linguistic facts are the *only* fact-like expressions of facts ready to our mouths and hands (cf. TLP 3.314ff.), only linguistic facts have the capacity to represent. This given, it is *language* that offers the means to fit sayings with the world of thoughts, sayings and other facts – thus justifying a *purely* linguistic view of representation.

This last premise, however, is subject to criticism, above all to Wittgenstein's self-criticism (cf. PI; Mitchell 1988; Gebauer 2005). Although Wittgenstein never strictly defends nor confutes this premise, he certainly confronts it in the way that he realizes his 'pair of glasses' (i.e. his language-biased set of premises), but he fails in using his 'pair of glasses' to illuminate himself wearing it. This, however, needs further clarification.

In the *Philosophical Investigations* (*Philosophische Untersuchungen*; first published in 1953), Wittgenstein affirms that his view of representation requires (among other aspects) that it is possible to represent thoughts and mental images (cf. PI § 389). But the question then arises, how thoughts and mental images are given if only statements represent what thoughts and mental images convey and if what they convey can actually be *transformed* into statements. On that note, it is hard to get rid of some background intuitions, in particular the idea that what thoughts and mental images convey *makes sense* without *yet* being a set of propositions. Thus, it appears that linguistic reductionism does not withstand the alignment with intuitions pertaining to the *genesis* and *transformation* of thoughts and mental images (e.g. via "imagery", Mitchell 1988; or "spectatorship", Mitchell 1994, p. 13).

One could try to defend Wittgenstein's purely linguistic view of representation by affirming the claim that "[t]here are, indeed, things that cannot be put into words. They make themselves manifest. They are what is mystical" (TLP 6.522). But are *all* things that cannot be put into words really mystical in this sense? And what follows from that? There is something deeply troubling about the exclusion of 'mystic' things and that is "particularly the epistemological notion of mental imagery, the temptation to think of perception, cognition, thought, and even language as mechanisms for recording, storing, and retrieving pictorial representations" (Mitchell 1988, p. 364). Thus, apparently, there are some 'only-shown-facts' in the world that pretend quite successfully to be mental and practical conditions or by-products of linguistic representations, although they cannot represent or be represented within this framework.

But this challenges the whole framework (cf. PI I § 115): since either it is not possible to represent all that is fact, including the conditions and effects of representation, which would make representation itself a partial mystery, or there are other, non-linguistic means of representation that can do the job. And from this point on, it stands to reason that, if we hold onto the mysteries, we accept too narrow a concept of representation, which is a bad deal compared to the tempting alternative of a more fine-grained, not entirely language-based view of representation.

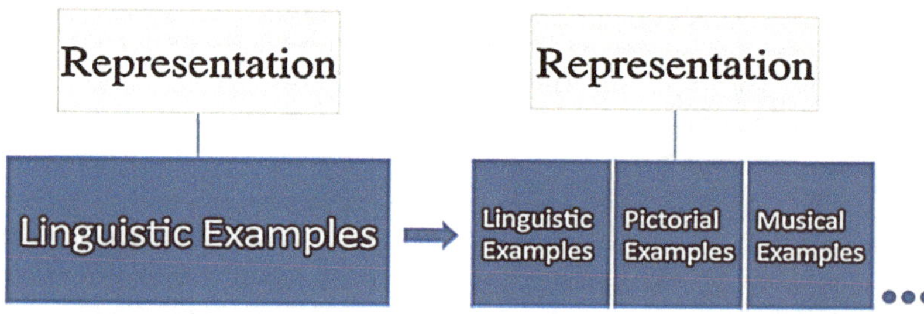

Fig. 3: Motivator [I]: Revision of the purely linguistic view of representation

Therefore, I will sum up the first motivator for further analyses of pictorial means as follows (cf. Fig. 3): the purely linguistic view of representation apparently fails to illuminate its own departure point, namely that *all* representations are language-like. Even from a purely linguistic view of representation it seems easier to give in to the idea that language is not the ultimate *tertium comparationis* of all representations (cf. Mitchell 1988, p. 364). And this is why Wittgenstein works (more or less unwittingly) as a great motivator for the analysis of pictorial means, because he pushed the linguistic view of representation to its limits, whereby these limits become ever more apparent.[18] From this point on, the ground is paved for an in-depth analysis of pictorial notions and means of representation, and also for synergies with current research on *scientific representation*, which is the subject of the next paragraph.

18 Wittgenstein's writings are a source of friction for many authors who work on pictorial means. Among those who draw on his notes are N. Goodman (1976), R. Wollheim (1985), K. Walton (1990) and O. Scholz (2004).

2.3.2 The role of picture-like means in recent accounts of science

In the previous section, I argued that it makes sense to devote attention to pictorial means after a period[19] of language-biased approaches, because the view thus extended converges better than its purely linguistic alternative with the range of aspects associated with the term 'representation' – a key term, in fact, since "current philosophies of science of all stripes agree with a characterization of science as an activity aiming at representing parts of the world" (Frigg 2010, p. 98). The present paragraph now continues with the task of motivating the idea and intention of this thesis and provides specific support to its *science-relatedness* for this purpose. This will cost some effort, as it is not possible to profile the relevant shifts in the philosophy of science using just one approach as a case in point (which was possible in the previous paragraph; cf. 2.3.1). Therefore, the second motivator divides into a whole cluster of motivating factors.

A good point to start with is the observation by G. Boehm (1994) according to which formerly language-focused disciplines have lately shown a growing interest in pictorial means (Boehm 1994, p. 13ff.). This growing interest clearly applies to the philosophy of science as well, which is striking insofar as philosophers of science neglected pictorial means for quite a while (cf. Kötter 2007, p. 161).

The remarkable change from indifference to interest suggests that something happened on the conceptual level before the new interest in pictorial means became manifest on the institutional level.[20] So, in order to provide a motivator for a *science-related* analysis of pictorial means, it seems appropriate to point out what changed in conceptual terms so that these means (and dedicated analyses) started to play a role at all in the philosophy of science. In order to shed light on the corresponding shifts and changes, a brief survey of recent debates in philosophy of science is required; for this survey assembles all the conceptual and historical information necessary in order to launch a *state-of-art argument* in favour of the science-relatedness of this work.

---

19 I refrain from a temporal pinpointing of this period, as there has been work on pictorial means *before* (Ch. S. Peirce; L. Fleck), *while* (E. Hutten) and, of course, *after* (B. Baigrie; etc.) R. Rorty (1967) stated the 'linguistic turn'. Thus, not all authors shared the limited focus on language at a certain time.

20 J. Elkins (2003) provided an overview of new research designs and publishing organs that operate under the name (resp. in conceptual nearness) to 'visual studies'.

In fact, there is not much room for variations of the storyline when surveying the research landscape in the philosophy of science. It has become a truism that the neglect of pictorial means in the philosophy of science is (here as well) due to *language-focused accounts* of science, which occurred in close personal and conceptual proximity to the linguistic view of representation (as outlined in the last section; cf. 2.3.1).

The 20th century started with influential approaches in this respect when considering the debates as coined by R. Carnap, H. Hahn, O. Neurath, H. Reichenbach, M. Schlick and others. Neurath might be an exception in this array, since he invented the *pictogram* (cf. Neurath 1991; cf. Fig. 4). But apart from him, no one has invested any effort in doing research on pictorial means. And Neurath himself was thinking of informational and educational rather than investigatory merits when he developed schemes for the use of pictograms (cf. Hartmann 2006, p. 47).

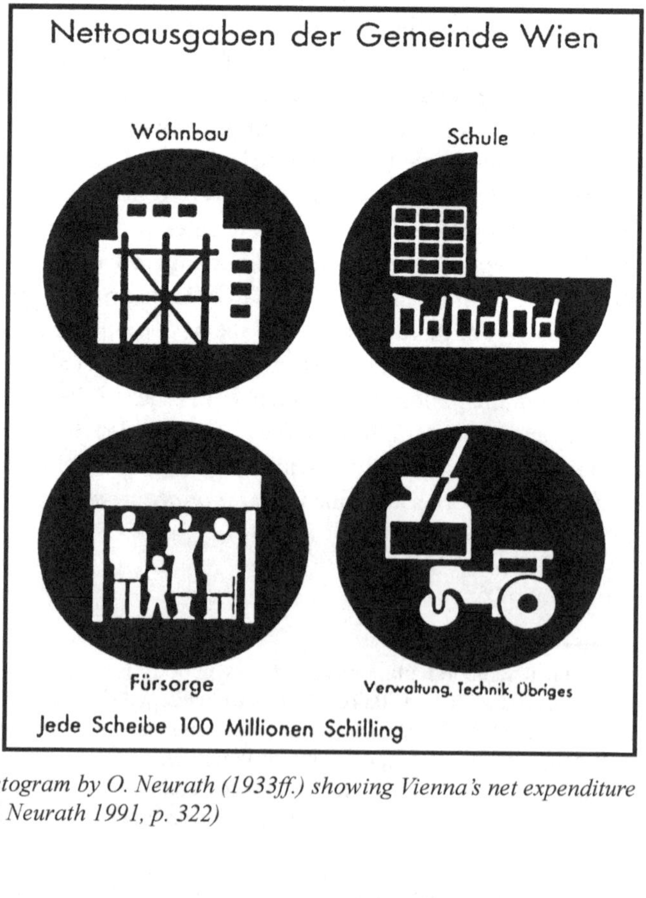

Fig. 4: Pictogram by O. Neurath (1933ff.) showing Vienna's net expenditure (cf. Neurath 1991, p. 322)

For this reason, it seems safe to affirm that the so-called 'logical empiricists' (cf. Reichenbach 1938, pref.) turned a blind eye to pictorial means. Yet the notion 'blind eye' is chosen with care, since, in all fairness, the logical empiricists did not explicitly debate the importance of pictorial means. They abstracted from pictorial means without an extra remark in the course of examining *language* as "the natural form of knowledge" (Reichenbach 1938, p. 17) and *logic* as the tool to master formal deficits of *ordinary language* (cf. Stegmüller 1969, p. 347f.). Thus, the logical empiricists set their focus on language *from the beginning on*, spurred by the wish for order and agility in what was initially set to be the realm of epistemology, namely "thinking formulated in language" (Reichenbach 1938, p. 16; cf. also Baird 2004, p. 1). The increased importance of language in this (virtually engineering) spirit was in turn motivated by the idea of providing a common foundation for the sciences, which faced two challenges at the beginning of the 20th century, namely (1) the *contamination by speculative approaches* and (2) the *call for a unified science* under the new paradigm of the *empirical* sciences. The idea was to tackle these challenges by uniting both "the empiricist conception of [experience-based] science and the formalist conception of logic" (Reichenbach 1938, preface). With this in mind, the attention turned to language, which seemed to be a tool that serves *general* and possibly *experience-related* tasks (e.g. description) under *self-imposed rules* and, thus, was seen as an apt candidate for a global methodology (cf. Carnap 1931, Hahn 1933).

However, by focusing on language and logic as constitutive instances of good science, other means that shape and further the processes in laboratories and research facilities faded into the background, especially those that did not comply straightforwardly with the items (terms, sentences) and rule sets (grammar, logic) under consideration, such as pictorial means. But, again, the point is that the logical empiricists did not explicitly tackle the question whether pictorial means comply with these items and rule sets or not. They simply premised that all relevant information could be broken down into verbal statements (cf. Carnap 1928 § 10ff.) in contrast to which pictorial means "can't help but [appear more] specific" (cf. van Fraassen 2008, p. 36f.) or at least different in kind. This admittedly rather psychological explanation for the reduction of propositions to logically treatable sentences finds further support with regard to the conceptual presumptions associated with logical empiricism. The fact that pictorial means actually remained in the background, even though they occurred in proximity to language-like and science-like procedures (such as representation), was further guaranteed by the standards of *epistemological relevance* as proposed by the logical empiricists. H. Reichenbach (1938), in particular, filtered epistemologically relevant and irrelevant information to the disadvantage of pictorial means (cf. Bailer-Jones 2009, p. 92). The respective filter is known as the distinction between the 'context of discovery' and the 'context of justification' to which epistemology is confined.[21] Accordingly, epistemology is *not* concerned with the circumstances under which belief sets are generated, but with the methods that help to constitute and adjust the logical relations between a theory and factual judgments, "independent of the man who found the theory" (Reichenbach 1938, p. 382; also Carnap 1928, 1984). K. Popper (1959), who was not in line with all the points of logical empiricism, also agreed with that belief. He affirmed that "[t]he initial stage, the act of conceiving or inventing a theory, seems to me neither to call for logical analysis nor to be susceptible of it" (Popper 1959, p. 31). Thus Reichenbach's (1938) distinction conveys a central premise of logical empiricism according to which it is sensible to detach psychology, which tackles empirical (esp. cognitive and habitual) conditions of science, from epistemology, which specifies conditions of linguistic and logical strength (cf. Neurath 1933, p. 8).

From this perspective, however, an epistemologically motivated analysis of pictorial means would find a sudden end here, as the logical empiricists refrain from investigating any other means apart from language and logic and any preparatory step of adapting the syntactic cores of propositions in order to establish their denotational and other (e.g. regulatory) function (cf. Giere 1996, p. 272). So, in fact, the focus was not only restricted to propositions, but to *verbally phrased, syntactically refined* and already *interpreted* propositions,[22] which were identified as the representative and representational essence of scientifically acquired

---

21 P. Hoyningen-Huene (2002) revealed five slightly different ways of giving sense to this distinction: (1) The 'context of discovery' and the 'context of justification' refer to different research stages that follow a chronological order; (2) The 'context of discovery' sheds light on research as a historical process and the 'context of justification' refers to methods of acquiring knowledge; (3) The 'context of discovery' is the realm of empirical studies, whereas the 'context of justification' is the realm of logical reasoning; (4) The 'context of justification' refers to the domain of the philosophy of science, but not so the 'context of discovery'; (5) They refer to different questions such as *Is a statement testable?* ('context of justification') or *What happened one after the other in a research process?* ('context of discovery').

22 B. C. van Fraassen (1991) clarified that propositions do not necessarily correspond with verbalized propositions: "That we cannot convey information, or say what a theory entails without using language, does not imply that – after all, we cannot say what anything is without language" (van Fraassen 1991, p. 5). This remark helps shed light on a possible stipulation underlying the architecture of logical empiricism, namely the synonymous use of '(verbalised) sentence' and 'proposition'. Only O. Neurath

knowledge, which is why this approach became known as the 'syntactic view of theories' (cf. Frigg & Hartmann 2006; a follow-up approach was provided, e.g., by Hempel 1965).

However, the protecting wall thus established against the necessity to consider non-linguistic means has become feeble in the meantime, as distinguished arguments have been brought up against logical empiricism, including the 'syntactic view' (for a historical overview, cf. e.g. Bailer-Jones 1999).

One of the weak points of logical empiricism (as outlined before) was undoubtedly the idea of establishing a meta-methodology that helps demarcate and conduct all scientific endeavours synchronically and diachronically by virtue of a template in basic reasoning. Among those who tackled this mission are (in alphabetical order): Feyerabend (2010), Fleck (1980), Hanson (1958), Kuhn (1996), Lakatos (1978), Laudan (1977), Polanyi (1958) and Toulmin (1972). These authors drew attention to the *historical dimension* of science claiming that science is – as other human activities – embedded in various technical, economical and social contexts that have an impact (perhaps more than logic) on how things develop in the realms of science. Hence, the authors mentioned suspended first of all the *normative* agenda of logical empiricism in favour of descriptive, context-related and tentative accounts of science, which shed light on the development of theories and shifts of "fashion" (Frigg 2003, p. 10) from historically attentive viewpoints.

But this reference to the historical variance of science was not the only reply to logical empiricism. While other authors remained lenient with ahistorical broad-spectrum accounts of science, they disagreed with the notion that sentences, grammatical ordinances and first order logic capture all functional components of that particular device – traditionally referred to as a 'theory' – that supports the basic tasks of science, namely to represent, explain or predict phenomena. Among those who joined in opposition to this view are (in alphabetical order): Cartwright (1999), van Fraassen (1980, 2002), French & Ladyman (1999), Frigg (2003, 2006, 2010), Giere (1988, 1999, 2004), Harré (1970), Hesse (1966, 1980), Hughes (1997), Morgan & Morrison (1999), Moulines (2002), Psillos (1999), Shapere (1984), Stegmüller (1976), Suppe (1989), Suppes (1969), etc. This sheer amount of dissenting votes against the 'syntactic view of theories' suggests how difficult it is to profile the counter-movement(s) without blurring individual differences.

One thing these authors certainly had in common was the more careful use of the term 'theory'; for the term 'theory' had a syntactic overtone in the ears of these

(1991) makes an exception. He specifies verbalised propositions ("*wortsprachliche Aussagen*") as a sub-kind of propositions (cf. Neurath 1991, p. 601).

opponents. Moreover, due to its broad range of application, which rendered the concept addressed indistinct, as U. Moulines (2002) stated: "[…] an adequate formal representation of the intuitive notion of a theory leads to at least three different *explicantes*. […] On the inferior (simpler) level, we encounter *theory-elements* [e.g. Newton's theory of gravitation]; on the medium level, we find *theory-nets* [e.g. Newtonian particle mechanics]; and the most complex units are called *theory-holons* [e.g. classical mechanics]" (p. 4).[23] Needless to say, this all-inclusive use of the term 'theory' was not regarded as helpful in operational respects, as it requires further specifications whenever it is referred to. And so, in response to the terminological ambiguity and conceptual indistinctness, the term 'model' found acceptance – either in addition to (e.g. Bailer-Jones 2009) or as a substitute for (e.g. Suppe 1977) the term 'theory'; since the term 'model' suited the terminological and conceptual intuitions that incited the revolt against the 'syntactic view of theories'.

The category of a 'model' seemed promising and more appropriate to many authors, since a 'model' "(a) is not supposed to cover *all* aspects of the empirical domain depicted by it, not even all of those considered important for some reason or other; (b) the 'model' is [also] not supposed to provide the ultimate truth about the domain in question, whatever this might be; [besides,] (c) the acceptance of the 'model' allows for successive revisions, refinements, suppressions, additions, etc., while maintaining its 'essentials' throughout all these changes" (Moulines 2002, p. 1f.). On that note, many agreed that not all what was formerly classified as a 'theory' actually bears all defining features (e.g. the universality) associated with this notion.

Complementary to this observation, which met the claim for more differentiated accounts of the conceptual devices employed in science, the diversity of the newly discussed devices also attracted attention: "Models can range from being objects, such as a toy aeroplane, to being theoretical, abstract entities, such as the Standard Model of the structure of matter and its fundamental particles" (Bailer-Jones 2002, p. 109). This diversity offered a further point of comparison in regard to its methodical relevance: "Models by their very constitution are applied to empirical phenomena, whereas theories are not" (cf. Bailer-Jones 2009, p. 3; cf. also Cartwright 1983). Thus, on the one hand, models were thought to adapt to the research objects to which they are 'applied'. On the other hand, insofar as they are not congruent with the range of research objects, they can almost be considered to be theories, and insofar as the concept of a theory was the predecessor from which the concept of a model was derived (cf. Falkenburg 1997, p. 28). Subsequently, the idea arose

---

23 The examples added in square brackets are taken from the same author (cf. Moulines 2002, p. 4).

(and has remained fixed ever since) that the newly approved concept of a model smoothes a not yet considered *gap* at some point *between* the most general and the case-variant frameworks employed in science. But, as one might expect, there are differences in how this gap is conceptualised and how models are fitted into it.

R. Frigg (2010), for instance, warns against formal accounts that seek to absorb models into algebraic concepts by defining them as set-theoretic structures[24] (cf. e.g. French & Ladyman 1999; van Fraassen 1980, 2002; Suppes 1957). Frigg contradicts such approaches, saying that reality is different, for scientists never communicate models as pure structures, "[because] even perfectly spherical planets are taken to have mass, populations are taken to consist of rabbits and foxes, etc." (Frigg 2010, p. 104). Accordingly, practice shows that models are semantically *specified* and thus treated differently than pure mathematical structures. Frigg (2010) furthermore argues that models are *inescapably* concrete in that sense, as they get disintegrate and become useless when one strips off their corona of attributions and interpretations (cf. Frigg 2010, p. 105ff.). So, Frigg (2010) affirms and further substantiates what others (e.g. Hutten 1956) put forth initially, namely that the context of use is significant for the otherwise underdetermined profile of theories or models. This assertion went along with the claim for considering all functional conditions – particularly the *semantics* – of conceptual devices and this, after all, became the common stance of the opponents of the 'syntactic view of theories' despite their discords concerning reductive vs. less reductive and formal vs. less formal approaches towards models (cf. Moulines 2002, p. 4).

But, as we can expect, there is no consent about the question of *what exactly* belongs to the 'semantic' profile of a theory or model and how this information can be extracted. The so-called 'constructivists' among the critics of the language-biased view of theories (in alphabetical order: Collins 2009; Knorr-Cetina 1990, 2001; Latour & Woolgar 1986; Latour 1987, 1996; Lynch 1985, Rouse 1996, etc.) seem to have the most particularistic view of the limits and prospects of profiling models:

> "Theoretical models often embody conditions that are unrealizable (sometimes in principle and sometimes as a matter of practical constraints on materiality or funds), whereas the actual construction of experimental systems often involves levels of specificity and local contingency which do not readily fit into models. Perhaps more

24 A 'structure' is defined "as a composite entity consisting of a non-empty set *U* of individuals called the domain (or universe) of the structure S and a non-empty indexed set R of relations on *U*" (Frigg 2010, p. 104). A 'structure' thus only defines how many objects there are (disregarding from their nature) and between which objects a relation holds (disregarding from their nature) (cf. Frigg 2010, p. 104).

> important, theory and experiment are developed in response to many considerations other than just their mutual articulation. [...] Thus we should see models and microworlds not as directed toward the *telos* of a unified practice of material and symbolic abstraction but as more dispersed and centrifugal engagements with the world" (Rouse 1996, p. 130).

In this spirit, the constructivists argue that scientific practices are deeply entrenched in the contexts in which they occur, because "understanding involves explicit beliefs and hypotheses, [and] these can only be meaningful in specific contexts and against a background of shared practices" (Dreyfus 1980, p. 7). Henceforth, a "tradition of philosophy of science arose that emphasized the importance of scientific practice to philosophical analysis" (Frigg 2010, p. 98; cf. also Morgan & Morrison 1999).

In fact, I got the impression that the majority of the critics of the 'syntactic view of theories' agree on shifting the focus to the realms of scientific practice (cf. also Frigg 2010, p. 98), exactly because it turned out that "understanding science" (cf. Rouse 1996) works best by virtue of low-scale approaches, which are the "touchstone of philosophical analysis" (Frigg 2010, p. 132), insofar as they get in touch with the practices to which assertions about science are fitted and by which the reflection on science is eventually motivated. Accordingly, one can find support for case-related studies throughout the spectrum of contributions to the debate – in formal (cf. e.g. Suppes 1957) as well as in less formal (cf. e.g. Morgan & Morrison 2009) accounts.

The specific contribution of constructivism is extreme caution concerning the generalisation of findings and, apart from that, a self-application of the "understanding science" agenda (cf. Rouse 1996); because, in the end, "theorizing [e.g. about scientific practice] is as much a practice as any other aspect of scientific work" (Rouse 1996, p. 127). Accordingly, not only science is claimed to be subject to historically variant rules of practice, but also philosophising on science itself; even more as the corresponding agenda shifted from regimenting scientific theory (as propagated by the logical empiricists) to illuminating variable scientific practices, a shift to which some refer to as the 'pragmatic turn'[25] (cf. e.g. Warnke 2002, p. 3; Pickering 1992).

---

25 The 'pragmatic turn' has two terminological relatives, the 'empirical turn' and the 'practical turn'. The 'empirical turn' was announced in the philosophy of technology (cf. Kroes & Meijers 2001; Achterhuis 2001), and the 'practical turn' denotes research on practical apprehensions of the world in epistemological terms (cf. Stern 2003). Hence, philosophy of science is not the only domain that turned to case-related analyses.

Having thus completed the journey through the research landscape, it should be clear now why *non-linguistic means* (such as *structures* [e.g. van Fraassen 1980] or *pictorial means* [e.g. Perini 2002ff.]) became the focus of attention and why it is, nevertheless, not trivial to reveal whether they are functional parts of the scientific acquisition of knowledge or not: The reason why this is difficult to decide is that ever since the rejection of the 'syntactic view of theories', the terms 'model' and 'theory' refer to conceptual *placeholders* in the first place, since their former standard analysis has been removed. On the one hand, this now allows non-linguistic means to play a role as *potential* components of models or theories. On the other hand, as indicated before, this role can hardly be profiled irrespective of concrete examples, because, in the terms of E. Hutten (1956), "[it] is obviously best to follow the scientists here as closely as possible, [in order to] avoid forcing science into a preconceived scheme" (Hutten 1956, p. 81). As a result, we face a variety of approaches towards theories or models (cf. Goodman 1976, p. 171) which correlate with scientific practices perhaps in the same way like models correlate with their references. Accordingly, Moulines (2002), for instance, has introduced his 'structuralist' account of models as a "particular program for modelling empirical science" (Moulines 2002, p. 1). In the same spirit, Frigg (2003) has chosen the title "Re-Presenting Scientific Representation" for his thesis on scientific models and the model-target relation. On that score, the notorious difficulty of providing a consensual working definition for the terms 'model' and 'theory' makes perfect sense within the modelling-focused and – for themselves *also* – model- and case-dependent accounts of science.

When regarding the motivational task of this chapter, we can conclude that the *modus operandi* in recent philosophies of science supports all the three starting points of a *case-* and *science-related* analysis of *pictorial means*. The respective motivators can be thus listed under three points (cf. Fig. 5):

(1) Recent approaches refrain from establishing a methodology for the empirical sciences and strive for "understanding science" (Rouse 1996) instead, that is, they try to grasp and illuminate the means that aid the scientific acquisition of knowledge.
(2) In close connection with the first point, current approaches prefer case-related to over-generalized conceptions of science.
(3) Finally, instead of pre-selecting the scope of epistemologically relevant instances to logically treatable sentences, "the role of language (and especially syntax or questions of axiomatization) is resolutely de-emphasized" (van Fraassen 1991, p. 7) in recent approaches, where attention is drawn to all sorts

of functional components (including pictorial means) that contribute to the accomplishment of scientific purposes.

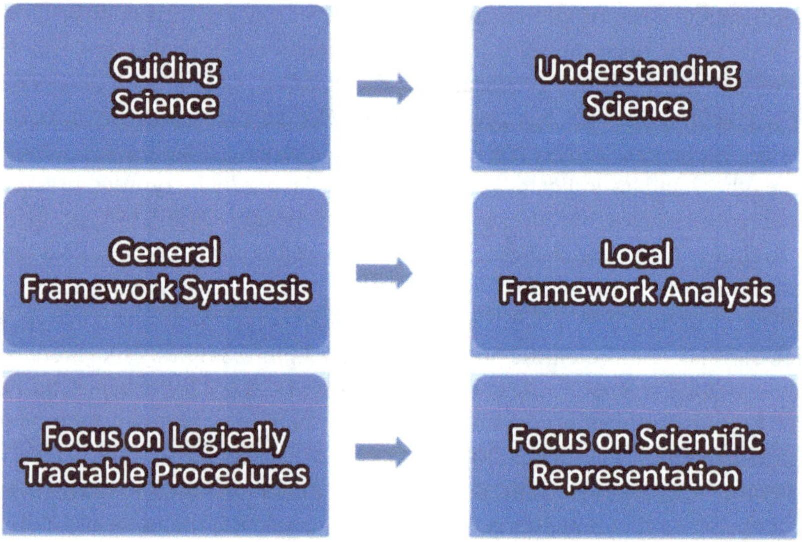

Fig. 5: Motivator [II]: Focal shifts in the philosophy of science

Now, having thus examined the science- and case-relatedness of this thesis against the background of the focal shifts in 20th century's philosophy of science, one possible objection remains to be tackled, namely why there are any further research efforts needed in this field, if it is accepted that the deficits of logical empiricism have been successfully pinpointed. In this context, it is important to stress, first of all, that the rejection of logical empiricism resulted in a withdrawal rather than a corroboration of clearing pretensions, which means that the concept of science is on the table again (cf. Gaycken 2005, p. 1).

Yet, even so, some legacies of logical empiricism have still not been as thoroughly discussed as the 'syntactic view of theories'. Non-linguistic means – and pictorial means in particular – are, for instance, still opposed to linguistic means when it comes to encapsulating and comparing their epistemic function. In the introduction to the philosophy of science by H. Poser (2001), for example, one finds the premise that scientific knowledge occurs in *linguistic* form (cf. Poser 2001, p. 28ff.; cf. 2.2.1.3). In this case, what might serve to restrict the scope of consideration is not even tackled in the BMBF exploration study on future research demands in the field of scientific visualization presented by M. Hessler (2004). Here, one finds the unquestioned premise that "the visual acquisition and presentation of knowledge is different from the discursive respectively formal-logical one"[26] (cf. Hessler 2004, p. 16). In contrast, a number of works have been published recently that focus on the epistemic relevance and role of pictorial means, pointing to a more careful stance on this issue (cf. 2.3.3; 4.1).[27] Moreover, to add weight to this argument, the *Forschungsinstitut für Philosophie Hannover* dedicated the 2010 research award to the question: "Can images be arguments?" ("*Können Bilder Argumente sein?*"). This can be seen as a further indicator for the fact that the community expects a review of habitual premises, especially regarding the epistemological status of pictorial means which recent approaches are not capable of illuminating (as will be shown in chap. 4.1).

But, before coming to the discussion of current frameworks, a third motivator in support of the idea and design of this thesis will be introduced, because, so far, support has mainly been given to research on non-linguistic means *in general*. The third motivator is thus meant to provide some back up for the particular focus on *pictorial* means using a *science-history* argument.

2.3.3 Techno-science and the emergence of scientific visualisation

The previous paragraphs revealed that pictorial means have gained attention along with the revision of language-bounded views of science (cf. 2.3.2) and (scientific) representation (cf. 2.3.1). The ongoing revisions are relevant to a plethora of arguments, pointing to ill-founded premises in connection with the linguistic reductionism. Criticism in this direction comes particularly from the proponents of the recently prevailing 'semantic view of theories' (cf. 2.3.2), which takes language as *one* ingredient of the scientific purchase of knowledge *among others*, following the intuition: "Science is not only language"[28] (Breidbach 2005, p. 172). However,

26 Translation of the quote: „die visuelle Wissensproduktion und -darstellung unterscheidet sich von diskursiven bzw. formal-logischen" (Hessler 2004, p. 16).

27 In addition to the methodology-focused literature surveys (cf. 2.3), a more concept-focused review is provided below, where it supports the choice of reference models and frameworks (cf. 4.1).

28 Translation of the following quote: „Wissenschaft ist nicht nur Sprache" (Breidbach 2005, p. 172). That science is not *only* language does, by the way, not mean that it is

as was said before, this intuition motivates research on all kinds of *non-linguistic* means – that is, not *only* on pictorial means – for it opens the former list of characteristic instances without clearly indicating what else or what instead marks the theory- or model-building. Therefore, it is time to substantiate this intuition with special regards to concrete concerns and hopes regarding the employment of *pictorial* means; this will provide for the third (this time historically warranted) motivator of this analysis.

When considering related publications to this theme it becomes apparent, first of all, that the history of science – although of younger age for itself – is acquainted with numerous uses and usages of pictorial means throughout the ages and disciplines. Accordingly, there are already a good many studies, which locate pictorial means in regional and systematic proximity to what they refer to as a 'scientific context' (see below). The range of studies increases even more when choosing the broadest possible notion of 'scientific' – that is, when taking all studies into account that investigate the purchase, distribution and storage of knowledge in whatever 'institutional' context (i.e. also in monasteries, private academies, intelligence services, etc.). M. Carruthers (1998, 2002), for instance, examined picture-based *mnemonic techniques* that she found elaborated in bookmaking and illumination (esp. of sacral texts) dating from the Middle Ages. Carruthers thus draws attention to early examples that suggest how fiendishly difficult it is to draw a line between linguistic and pictorial procedures when it comes to illuminating the "memory-making" (Carruthers 2002, p. 2) and the "making of ideas" (Carruthers 1998, p. 7), since linguistic and pictorial measures occur in operational proximity, that is, they are found justified with reference to at least one common target (here: "composing thoughts and memories", Carruthers 1998, p. 2).

Such findings (e.g. Carruthers 1998, 2002) brought about further research on pictorial means in familiar contexts. Yet it is obvious that corresponding research is mostly conducted from historical points of view (in alphabetical order: Adelmann et al. 2009, Baigrie 1996, Breidbach 2005, Bredekamp 2000, 2003ff., 2008, Daston & Galison 1992, 2007, Elkins 2007, Gall 2007, Gugerli 2002, Hentschel 2000, Hessler 2006, Huber 2001, Kemp 1990, 2000, Kötter & Inhetveen 1996, Panofsky 1957, Rheinberger 1997, 2001, 2005; Robin 1992, Vogelgsang 2011, etc.). The authors and editors listed allow detailed insight into testimonies of pictorial procedures to which they apply scientific descriptors and classifiers (such as

---

not *mainly* language. K. Sachs-Hombach 2009) rightly annotates that pictorial means rarely occur on their very own in the domains of science, whereas texts are not always found in pictorial neighbourhoods (cf. Sachs-Hombach 2009, p. 9).

'observation' and 'reasoning'). Altogether, the authors cover a considerable period of time (without suggesting a continuous genealogy), starting with Aristotle's zoology (~ 400 BC), crossing Hippocrates's studies on human diseases (~ 400 BC), Galileo's sketches of the lunar surface (~1609), Descartes' geometrical analyses (1637), Darwin's study on emotions of man and animals (1872), reaching 20th century's data imaging – to mention only a tiny selection of the discussed examples. Other authors and editors contribute to the list by exploring pictorial measures throughout academia at a certain time and place, thereby unfolding a synchronic view (cf. Beyer & Lohoff 2005, Elkins 2007, Frankel 2002, etc.).

But before pictorial means (occasionally specified as 'scientific images'[29]) actually reached the stage and status of research objects in history and philosophy of science (intensively since the 1990's) their abundant use gave rise (among art-historical issues) to *anthropological* discussions (as launched, e.g., by Assmann 2009, Cassirer 2010, Jonas 1961, Langer 1979, etc.). The anthropological debate treats pictorial means as symbolic devices, which are referred to as species-specific developments that are possibly relevant to the distinction between human beings and other beings. H. Jonas (1961), for instance, introduced the term "*homo pictor*" in order to convey the idea of the picture-using – therefore *human* – species. This idea was, in fact, coined by H. Cassirer (2010), who specified humans as the "*animal symbolicum*", a notion that occasionally reached the realms of biology (cf. Bertalanffy 1968), where symbolic devices were also discussed as potential candidates for the *differentia specificam*, that is, for the very feature or set of features that distinguishes the human condition from that of other species.

And so, to cut short the inventory of publications: a cursory glance at the history-based literature reveals that there are already lively (albeit disparate) discussions on pictorial means, which influence situations, terms and concepts akin to the philosophy of science. However, the philosophy of science itself struggles in coping with the historically developed sources of information about pictorial practices in scientific or broader epistemic contexts (cf. 2.1.2; 4.1).

---

29 There are attempts (particularly found in the German literature) to establish a technical term for the pictorial means that were found in scientific contexts (cf. e.g. „*Wissenschaftsbilder*", Heßler 2006, or „*Bilder des Wissens*", Breidbach 2005). However, the suggested terms have not yet reached the status of technical terms. Some argue that these terms evoke the *wrong* idea that pictorial means provide context-invariant, ready-to-hand knowledge (cf. Kötter 2007, p. 164). I agree with this objection, and I would like to add that using the suggested terms as technical terms yields no significant economical or rhetorical benefit, since further specifications and explanations are required each time for the mentioned reason.

But before coming to review those approaches that contribute to the 'pathfinding' in the present philosophy of science, some further historical remarks on pictorial means are indispensible. Because, so far, one might have gotten the impression that pictorial means have found proponents at all times, up to the present day, which is *not* the case. In fact, the current prospering of a "very active and vital area of [picture-based] research, teaching, and development" (Nielson, Hagen & Müller 1997, p. xiii) easily masks the fact that there were (downright emotional) stances on the *disadvantages* and *risks* of the scientifically motivated utilisation of pictorial means. These opinions, however, can hardly be explained with an ever-flourishing iconoclasm in the tradition of Plato's downgrading of the art of painting (cf. *Politeia* 597b–598e; Mersch 2006a). Certainly one finds these opinions justified against the background of contemporary scientific processing (see below), which indicates a non-scholastic, self-directed process of judging and evaluation. Still, this issue has been avoided in recent literature (so I will for the most draw on the historical overviews provided by Bailer-Jones 2009 and Morgan & Morrison 2009), although the disapproval of pictorial means offers the *key* to a strong motivator for a philosophy of science perspective on the current role and relevance of pictorial means. The motivation arises particularly from the confrontation of the recent enthusiasm with the former reluctance, which reveals a remarkable liberalism among scientists at the present time (cf. Bredekamp 2008, p. 47). The liberalism manifests itself in the techno-scientific enhancement of picturing techniques starting from the late 70's on (as exposed below). It is even claimed in this context that "the for a long time resistant humanities [...] attend more and more to the use of pictorial means for epistemic or, at least, cognitive purposes so that, meanwhile, one speaks of a 'pictorialisation of the sciences'"[30] (Beyer & Lohoff 2005, p. 11). However, in order to transform these 'signs of the time' into a convenient motivator for the idea and design of this analysis, a detailed review of the arguments for and against the scientifically motivated employment of pictorial means is required in the following paragraphs.

On that note, the 19th century, in particular, offers valuable points to start with, also because current authors pick their earliest references from this period (cf. Bailer-Jones 2009, p. 21). What makes this period noteworthy is that "the reliability of visual thinking [...] came under heavy suspicion in the nineteenth century" (Giaquinto 2007, p. 3). This holds for the field of mathematics (studied,

30 Translation of the following quote (square brackets added): „die lange Zeit weitgehend bildresistenten Geisteswissenschaften [...] bedienen sich heute vermehrt der Bilder in epistemischen oder wenigstens kognitivem Interesse, so dass bereits von einer ‚Pictoralisierung der Wissenschaften' gesprochen wird" (Beyer & Lohoff 2005, p. 11).

for instance, by Giaquinto 2007) as well as for those fields that started to draw strongly on mathematics (particularly on algebra), like physics and other "Baconian science[s]" (Bailer-Jones 2009, p. 23).

And so, J.-L. Lagrange (1997), for instance, announced the following constraints in the preface to the first edition of the *Mécanique Analytique*:

> "No figures will be found in this work. The methods I present require neither constructions nor geometrical or mechanical arguments, but solely algebraic operations subject to a regular and uniform procedure"[31] (Lagrange 1997, pref.).

Apparently, Lagrange considered this statement important enough to place it in the preface of his treatise. He thus provided the treatise with a preliminary remark on the employed operative constraints, which, in his terms, exclude the *non-constrainable* handling of pictorial means in general. Pictorial means are thus treated as untameable, therefore obsolete, supplements for the reason that they do not correspond to algebraically described and justified operations.

However, pictorial means were not the only means that gained a "poor reputation" (Bailer-Jones 2009, p. 16) on that score. Models had the same fate, although it is difficult to judge from the available historical sources, whether models diminished the reputation of pictorial means, of which they may be components, or the other way round. It can certainly be said that the "well-accepted mathematical trend" (Bailer-Jones 2009, p. 23) at that time qualified 'formal' expressions and operations as scientific devices in the first place (cf. 2.3.2; 4.2.5.4; 4.3.2.3); everything else was widely deemed to be of minor, preparatory or post-procedural importance:

> "Many thought that the place of models could or should be taken by rigorous mathematical treatment and that this, although it may have emerged from models, would supersede them" (Bailer-Jones 2009, p. 32).
>
> "This trend gathered momentum in the early days of the twentieth century. Models, as understood in the nineteenth century as physical objects, were then at best thought to have temporary and heuristic advantages, being useful for cognitive and practical purposes, while a contrary widespread view was that they could even be eliminated in principle" (Bailer-Jones 2009, p. 32).

Hence, the discrimination of non-mathematical (esp. non-algebraic) means affected the status of both models and pictorial means; reportedly as both were taken

---

31 Translation by A. Boissonnade and V. N. Vagliente of the following quote: „On ne trouvera point de Figures dans cet Ouvrage. Les méthodes que j'y expose ne demandent ni constructions, ni raissonnemens géométriques ou méchaniques, mais seulement des opérations algébrique, assujetties à une marche régulière et uniforme" (J.-L. Lagrange 1853, pref.).

to have a material-bounded genesis and range of functions, which comes closer to artisanal work – especially to the manufacturing of "tangible representations" (Boltzmann 1911), such as installations for the "pushing and pulling of things" (Bailer-Jones 2009, p. 21; cf. e.g. Fig. 6) – than to pen and paper considerations.

The preference for formal operations and the consequent downgrading of non-formal procedures went on until the midst 20th century in the case of models, and until the late 20th century in the case of pictorial means respectively. And even some of the contemporaries of logical empiricism, who launched *post-empiricist* thoughts, like P. Duhem (1998), for instance, took a deprecatory stance on models and pictorial means. Accordingly, P. Duhem (1998), on the one hand, came up with an influential analysis of *scientific observation* that exposed the latter as *theory-loaded* (which became a serious challenge to naïve realism as well as to the correspondence theory of truth; cf. Hesse 1980, p. vii). On the other hand, in regard to models and pictorial means, Duhem wanted the academia to "keep [physicians] away from this useless work"[32] (Duhem 1998, p. 130). This remark was plainly construed by E. Mach, who noted that: "Duhem regards the model as well as the picture as parasitic excrescences"[33] (cf. Duhem 1998, preface, IV). With this, Duhem remained in line with the 19th century's spirit that welcomed "the mathematization of the empirical sciences" (Bailer-Jones 2009, p. 23) as opposed to the conventional mechanical modelling, to which the term 'modelling' was confined at that time.

Another programmatic campaign against pictorial means was launched in the 20th century, when a group of mathematicians, operating under the common pseudonym *Nicolas Bourbaki*,[34] entered the academic scene. When P. Cartier, a

---

32 Translation of the following quote [square brackets added] from the 1st edition in German language (1998): „die Physiker von dieser nutzlosen Arbeit abhalten" (Duhem 1998, p. 130). Due to the fact that Duhem rejected models and pictorial means, I did not mention him earlier in chap. 2.1.2, although he objects to some key conceptions of logical empiricism and could thus be added to the list of opponents (esp. of the hypothetico-deductive reasoning; cf. Carrier 2006, p. 43ff.).

33 Translation of the following quote from the 1st edition in German: „Das Modell wie das Bild hält Duhem für ein parasitäres Gewächs" (Duhem 1998, preface, IV). This does not include so-called *theoretical* or *abstract pictures*, for which he used the same noun (cf. Duhem 1998, p. 366). So did H. Hertz (1894), who coined an abstract notion of the picture, which was picked up by Wittgenstein, who in turn linked the concept of depiction (resp. representation) with the concept of formal correspondence (cf. 2.3.1).

34 *Nicolas Boubaki* was an author collective, which devoted itself to a new foundation of mathematics. According to M. Senechal (1998): "Bourbaki was born in Paris in 1935 when a small group of mathematicians at the École Normale Supérieure, dissatisfied with the courses they were teaching, decided to reformulate them" (Senechal

former member of Bourbaki, was interviewed about the hypotheses that guided this endeavour, and especially the associated rejection of pictorial means, he presented a distinction that harks back to the 19th century in terms of categorical rigour:

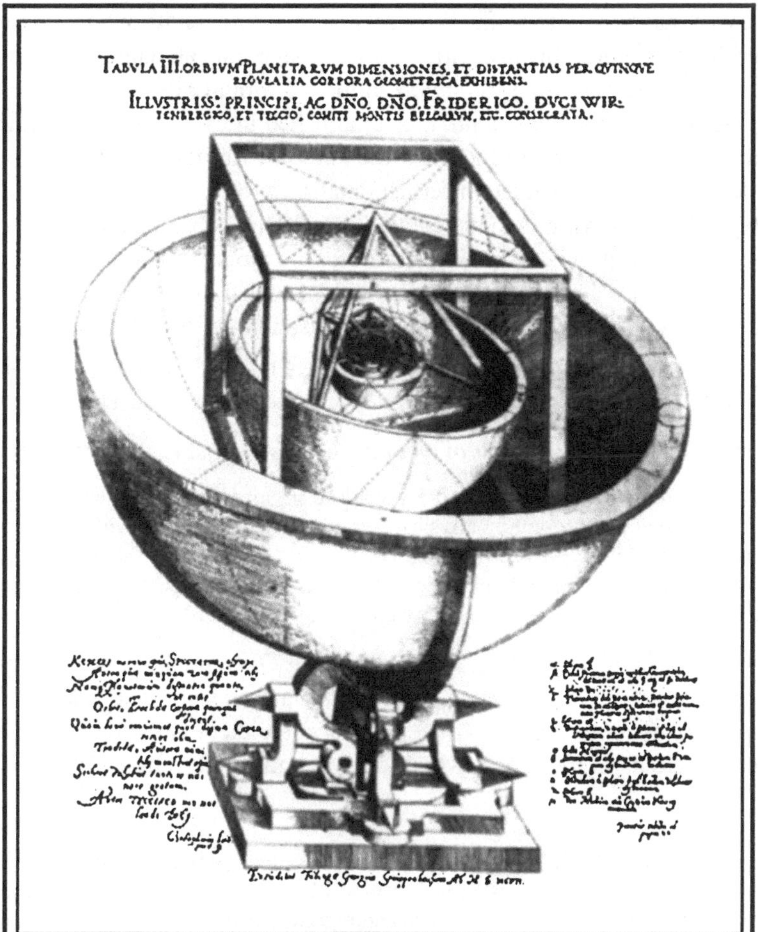

Fig. 6: Solar system modeled by Johannes Kepler, 1596 (cf. Kepler 1938)

1998, p. 22). The members of Bourbaki set common rules for their work (including anonymous publishing) and sought for a "total formalization and perfect rigor" (Senechal 1998, p. 22). Bourbaki is said to have a lasting impact on the way mathematics is conceived and conducted (cf. Senechal 1998, p. 22).

> "[T]here was the idea that there is an opposition between art and science. Art is fragile and mortal, because it appeals to emotions, to visual meaning, and to unstated analogies. [...] [S]o-called geometrical intuition was not reliable in proof" (Cartier 1997, p. 27).

From what has been said so far, there seem to be some points of agreement among the critics of pictorial means, which Cartier also points at, namely the belief that: *scientific* procedures are *by definition* attuned to formal rules and schemes of proceeding; and that the constraints thus established in execution and evaluation do not apply to pictorial procedures, for the latter are unruly or unreliable (why this should be so is not explained in further detail).

Beyond all doubts, it is to the credit of those who subscribed to the idea of formalisation that it proved extremely useful for the progress of science (for a historical overview, cf. Krämer 1988). Accordingly, O. Neurath (1991), to name but one, commented favourably on the idea of formalisation (cf. also 2.3.2):

> "The expression of a scientific result, particularly a formula, is meant to be as clear as possible. That is exactly its advantage against any other aspect"[35] (Neurath 1991, p. 601).

Neurath then exposed the limits of pictorial means against the so-defined benchmark:

> "Admittedly, not all aspects can be visualised in an adequate and acceptable way. When writing and presenting physical formulas, one often deals with very complex matters, for which there is no suitable visual representation. Only some rather simple relations can be visualised in an adequate and concise manner. The human potential to visualise [...] is rather limited"[36] (Neurath 1991, p. 602).

On the other hand, and this is significant about O. Neurath, he highly valued the employment of pictorial means for didactic, autodidactic and educational purposes, since "[f]rom a pedagogical point of view there is the danger of enslavement by verbal formulations, which can encumber the rational thinking due to

---

35 Translation of the following quote: „Die Aussage eines wissenschaftlichen Ergebnisses, besonders eine Formel, ist absichtlich so klar wie möglich gehalten. Gerade hierin besteht der Vorteil gegenüber einem anderen Gesichtspunkt" (Neurath 1991, p. 601).

36 Translation of the following quote: „[Es] muß offen anerkannt werden, daß nicht alle Zusammenhänge in geeigneter und angemessener Weise visuell dargestellt werden können. Beim Schreiben und Darstellen von Formeln in der Physik hat man es beispielsweise häufig mit sehr komplizierten Zusammenhängen zu tun, die auf diese Weise nicht passend dargestellt werden können. Nur gewisse relativ einfache Zusammenhänge können eindrucksvoller und knapper in visueller Form dargestellt werden. Die menschlichen Fähigkeiten zur Veranschaulichung [...] sind relativ begrenzt" (Neurath 1991, p. 602).

their stiffness"[37] (Neurath 1991, p. 601). Accordingly, despite a trend towards "mathematization" (cf. Bailer-Jones 2009, p. 25), non-mathematical procedures were expected to facilitate the individual work with formulas. Some expected even more in the empirical domains (cf. e.g. Marey 1878).

This also fits, for instance, W. Thomson's (aka Lord Kelvin) reaction on J. C. Maxwell's theory of the electro-magnetic field. Maxwell devised this theory by virtue of a pen and paper analogy between vortexes in a fluid and the electro-magnetic field (cf. Bailer-Jones 2009, p. 25ff.). As soon as he achieved the mathematical formulation of the electro-magnetic field, he jettisoned the preliminary drafts and published the formulation only (cf. Morgan & Morrison 2009, p. 33ff.). This move demanded an explicit instruction and rationale:

> "The substance here treated of must not be assumed to possess any of the properties of ordinary fluids except those of freedom of motion and resistance to compression. It is not even a hypothetical fluid, which is introduced to explain actual phenomena. It is merely a collection of imaginary properties which may be employed for establishing certain theorems in pure mathematics in a way more intelligible to many minds and more applicable to physical problems than that in which algebraic symbols alone are used" (Maxwell 1965, p. 160).

J. C. Maxwell thus connected a mathematical and mechanical approach in the sense that "a mechanical model of a phenomenon was mathematically described" (Bailer-Jones 2009, p. 25). As indicated above, Thomson reacted critically to this:

> "I never satisfy myself until I can make a mechanical model of a thing. If I can make a mechanical model I can understand it. As long as I cannot make a mechanical model all the way through I cannot understand; and that is why I cannot get the electro-magnetic theory. I firmly believe in an electro-magnetic theory of light, and that when we understand electricity and magnetism and light, we shall see them all together as part of a whole. But I want to understand light as well as I can without introducing things that we understand even less" (Thomson 1987, p. 206).

Accordingly, Thomson found himself prompted to justify the plea for a conventional mechanical model in addition to the "physical analogy" (Maxwell 1965, p. 156) that was offered by Maxwell. Morgan & Morrison (2009) provided a helpful exemplification of Thomson's desire:

---

37 Translation of the following quote: „Aber vom pädagogischen Gesichtspunkt besteht die ständige Gefahr einer Versklavung durch wortsprachliche Formulierungen, die durch ihre Starrheit das vernünftige Denken tatsächlich behindern können" (Neurath 1991, p. 601).

> "For example, when demonstrating the model of a molecule interacting with the aether Kelvin used a ball in a bowl of jelly and produced vibrations by moving the ball back and forth. For him this provided a more realistic account of the relation between matter and the aether because one could see how vibrations were produced and travelled through the medium. Kelvin claimed these models were demonstrative in a way that Maxwell's 'paper' models were not" (Margret & Morrison 2009, p. 37).

But Thomson's confidence in the (at his time more and more controversial) mechanical models was a matter of debate, especially since he needed to defend them against the growing trend of abstraction from sensual experience. This trend was inexorable, insofar as mechanical models turned out to be casually misleading or unrealisable (also, for example, in case of the electro-magnetic field theory; cf. Bailer-Jones 2009, p. 25). Still, it is to the credit of Thomson and others that "with the rise in the use of models in science, a debate started about what the merits of models were, if any" (Bailer-Jones 2009, p. 29). And – although it took a while – this debate finally became the centre of attention in the 20th century's philosophy of science, where the variety of modelling concepts increased tremendously, and where models became a most central issue (cf. 2.3.2).

However, as revealed before (cf. 2.3.2), there is a heated discussion about what exactly works as a model and how models relate to other operational and conceptual devices. While attention was first drawn to mathematical 'structures' in a more or less abstract sense (cf. e.g. van Fraassen 1980, 2002; Suppes 1957), the current discussion shows the first efforts to capture the *pictorial means* involved. The shift of attention correlates with the changes in research, since technology has flourished and brought (among other things) the digital computer, which has boosted mathematical procedures tremendously. This process was further accelerated from the late 1970's onwards, when computer-aided ad hoc visualisations became part of statistical, algebraic and topological methods (cf. Wong & Bergeron 1997, p. 4). From then on, the former ad-hoc measures were quickly transformed into a technological format. Meanwhile, the term 'scientific visualisation' has given its name to a distinct research agenda and associated industry. This involves *inter alia* a particular academic embedding, manuals (in alphabetical order: Brodlie et al. 1992, Gallagher 1995, Nielson, Hagen & Müller 1997, Schumann & Müller 2000, Spence 2007, Ward, Grinstein & Keim 2010, and more) and educational programs,[38] which introduce standard problems, methods and examples

38 Full study programs are currently offered at the University of Koblenz-Landau ('Computational Visual-istics') and at the Otto-von-Guericke University of Magdeburg ('Computational Visualistics'). Moreover, it is possible to specialise on

and which guide researchers in developing new research tasks based on previous achievements. This institutional embedding and appreciation has led to exchanges with established disciplines and has given rise to cross-sectional research fields, such as network analysis (cf. e.g. Krempel 2005; Schich 2009) and systems biology (cf. e.g. Schittler et al. 2010; cf. 3.2).

From the researchers' point of view, there are several factors that have fostered and still continue to further the growth of 'scientific visualisation' as a distinct area of expertise. First and foremost, there is the demand for solutions that facilitate the processing of huge amounts of information or data:[39]

> "The standard argument to promote scientific visualization is that today's researcher must consume ever higher volumes of numbers that gush, as if from a fire hose, out of super-computer simulations or high-powered scientific instruments. If researchers try to read the data, usually presented as vast numeric matrices, they will take in the information at snail's pace. If the information is rendered graphically, however, they can assimilate it at much faster rate" (Friedhoff & Kiley 1990).

Closely linked with the demand for efficient solutions in terms of information and data processing, there is the need for balancing strategies in regard to: "making the

---

'Computer Graphics' during a major in Informatics or Software Engineering, for example at the University of Stuttgart.

39 The terms 'data' and 'information' are used equivocally. P. C. Wong & R. D. Bergeron (1997), for example, consider 'information' as referring to the desired results of 'data' analyses (cf. Wong & Bergeron 1997, p. 5). 'Information' thus denotes properties of data sets, which are deemed relevant against the background of particular research interests (cf. Wong & Bergeron 1997, p. 7). Unfortunately, the authors refrain from defining 'data' itself; they might take it for granted that 'data' often denotes the numeric results of simulations or measurements. A. Luft & R. Kötter (1994), on the other hand, proposed a 'pragmatic conception' of 'information' and 'data'. Accordingly, the term 'information' refers to speech acts, which are undertaken in order to modify someone's state of knowledge (cf. Luft & Kötter 1994, p. 205), whereas the term 'data' refers to those elements of the speech acts, which are relevant for the receiver of the information (cf. Luft & Kötter 1994, p. 208ff.). For example: A person can inform herself by reading the display of a barometer, but only if she is acquainted with meteorological methods, she can gain data of possible use from this information (cf. Luft & Kötter 1994, p. 210). The difference is that Luft & Kötter therewith try to avoid a step backwards into naïve realism in the way that they emphasise that properties are functional parts of propositions and not of 'pure nature'. Hence, they insist that information is the raw material for the analysis of relevant implications, which is data. For further work on the terms, cf. e.g. Chen (2009).

mathematics applicable to physical problems, [and] making it intelligible to the mind" (Bailer-Jones 2009, p. 30):

> "The goal of scientific visualization is to promote a deeper level of understanding of the data under investigation and to foster new insight into the underlying processes, relying on the human's powerful ability to visualize" (Earnshaw 1992, p. 1).

The important aspect about the current ways of promoting computer-aided uses of picture-like means is a change in tenor: while pictorial means counted as loose additives in former times, they have proved to be so useful that now a whole research environment and associated technology has formed around the idea to explore and exploit their enabling potential (cf. 3.1f.). So this is the point where the pressure to revise former views of pictorial means in science becomes "inescapable" (Mitchell 1994, p. 16). For that reason, I suggest encapsulating the third motivator as follows (cf. Fig. 7): pictorial procedures have undergone a remarkable change of reputation in the scientific realms. Their status has shifted from being addressed as accompanying techniques to being addressed as an enabling technology. And this, after all, makes pictorial procedures worth analysing in regard to the questions previously exposed (cf. 2.2.1.3).

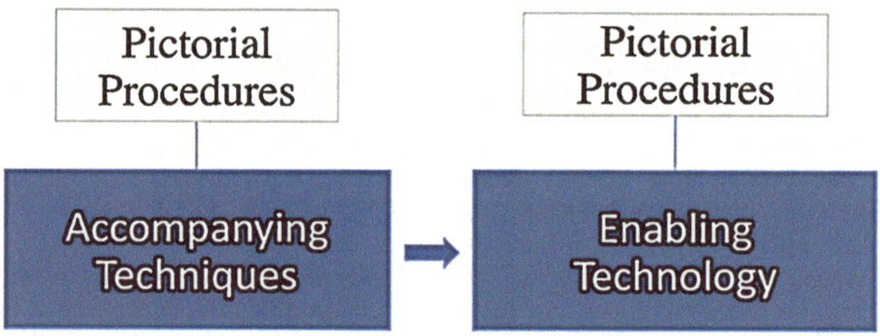

Fig. 7: Motivator [III]: Status shift concerning pictorial procedures in science

III Examples

Having defended the need for a revised view of pictorial means in science, I will proceed with the introduction of examples. Considering the line of argumentation, it makes sense to acquire some intuitions of actual usages of pictorial means before objecting to the notion that the latter are irrelevant or too unspecific for the scientific acquisition of knowledge or conceptual accounts thereof.

The idea now is, more precisely, to create an empirically supported reference base, which sets demands on the soundness and illuminative power of those accounts that seek to defend or reject the epistemic relevance and surpluses of picture-like means (reviewed in chap. 4.1ff.). In other words, the examples should not only work as motivators, but also as correctives of the positions discussed below. This is in accordance with the current mode of proceeding in the philosophy of science (as outlined above; cf. 2.3.2; 2.3.3), which is expected to be compatible – and should definitely not compete – with the self-conceptions of the domains under consideration.[40]

One simply needs to bear in mind that when focusing on a limited set of examples, the empirical significance of assertions thus supported is possibly limited, too; at any rate it can only be defended along the lines of the chosen examples. It goes without further saying that this is the reward from turning to case-related analyses. On the other hand, focusing on a limited set of examples also has specific advantages. First of all, one is free to pick examples from a distinguished source, which makes it possible to introduce the examples as new, original or prototypical in certain respects (e.g. from an author's or a research community's perspective).[41] Besides, working on non-standard examples (also from the *own* scientific community's perspective) may help detect problems of conceptualisation in the course of testing current descriptors and classifiers. And this is actually what this chapter

---

40 This means, e.g., drawing on accepted journal papers and personal discussions with the authors.

41 The examples are drawn from research projects that were conducted at the University of Stuttgart, where several contributions to scientific visualisation originate (from the Institute for Visualisation and Interactive Systems), and where advanced visual solutions are requested in the areas of simulation-based research (as boosted esp. in the research cluster 'Simulation Technology' [SimTech]).

is all about. Therefore, now, the task is to gather information and signs of problem awareness concerning the intricate relation between pictorial procedures and other scientifically motivated procedures in the course of a descriptive approach.

However, before leaping into the task, the idea of a *descriptive approach* calls for more of these adjoining remarks; because, given that the use of pictorial means implies a mode or measure of representation, and given that the use of verbal descriptors does so as well, the challenge here is to find a convenient way of *(re-) presenting representations* (cf. Frigg 2003). This problem came up beforehand (cf. 2.3.1). Its practical dimension, however, stands out all the more clearly when one tries to prepare an exemplary set of pictorial means, or that is to say, *testimonies* of pictorial means, for a closer inspection.

In more elaborate terms, one is compelled to ask: is it possible, and if so, how is it possible, to transfer pictorial means *as* pictorial means from one context (e.g. a journal paper) into the other (e.g. this thesis)? On a closer look, this question has a technical and a conceptual dimension: from the technical point of view, the question literally is, where to make the cut: below or above a figure's inscription? Close to or far from the centre of a figure? In this context, a conspicuous feature of the targeted entities becomes apparent, namely that their printed components – that is, what is actually presented to the eye *as* a 'figure' – are unframed or vaguely framed in the source documents (i.e. journal papers). Instead, there is white space between what is inscribed as a 'figure' and the surrounding text, offering no predefined cutting edges and leaving one to assume that figures are either not expected to be quoted, or – and that seems likely in the realms of scientific exchange – that it is up to the quoting authors to define the ends and edges as well as any additives on individual purpose. The conceptual concerns, on the other hand, are directly linked (if not precursory to) these technical quandaries. The question is, what must be transferred, also for example in terms of textual extracts, in order to preserve the nature and individual character of pictorial means? Apparently, and this can be guessed from the long history of aesthetics, the conditions of *completeness* (conceptual) and *completion* (technical) are far from being clear-cut in this case.

Before immediately deciding on these matters, it is worth going deeper into the question of why there is so much confusion at all. In this context, Wittgenstein's observations have proven quite instructive (cf. 2.3.1). Wittgenstein argued that representational means do not represent *their own mode of representation* (the related argument was briefly sketched in 2.3.1). Rather, there are further conditions that need to be met for instancing representations or for determining traces of them. For example, one needs to construe the *structures* of the representational and represented entities (TLP 2.15; 2.17) or their *contexts of use* (PI § 11; § 43).

Wittgenstein's intention aside, the community has reached a consensus stating that representational means always involve them being *things-in-use*, *things-of-use* or *things-for-use* rather than merely things (cf. also Schirra 1999, p. 104; Kjoerup 1978, p. 57). Hence, in order to characterise them properly, one needs to retrace their functional involvements in addition to the exposure and description of their sensorially tangible corpuses.

This fits the notion of *means* in general. For example, sticks with iron heads are not automatically treated as equivalent to 'hammers' or 'paperweights'. Only if certain sticks with iron heads are also associated with the beating, driving or loading of things, that is, with specific means-ends relations, are they conceived of as means (e.g. 'hammers').

Yet, while these are fairly simple examples, things become gradually more complex in terms of pictorial means, which might explain the initial perplexity concerning their accurate quotation; because, treating something *as* a pictorial means implies concurrent mechanical (framing, lighting, etc.) and interpretational measures (scanning, describing, classifying, etc.), not only, but also when one tries to infer to former usages of such means, pursuing this whole complex behind the occurrence of material traces. And this leads eventually to this point of friction: representing pictorial means requires inferences to entangled means-ends relations across different methodological domains (e.g. pictorial and scientific modes of proceeding), while not all factors that guide the inferences can be found documented *by* or *in* the very means under study. Hence, picture-like means do not represent the experiences and associations they trigger.

This compound of challenges is certainly not new to historians of art or science. Still, it puts the foundations of this thesis to a grave test, since, apart from the need to navigate through technical and conceptual quandaries, the question is: if we (scientists and layman alike) reconstruct pictorial procedures *verbally*, and if we conceive of pictorial means in terms of the whole of such a construal, do we then not somewhat accidentally support the notion that pictorial means are transformable into, if not genuinely present as, *verbalised* propositions?

This thesis certainly demands a negative reply, as it propounds a categorical difference between linguistic and pictorial means of representation (cf. 2.3.1) and, in this context, that non-verbal (esp. pictorial) means are used *as* non-verbal (esp. pictorial) means in the realms of science. The premises that are put to the test in this thesis thus conflict with a linguistic view of pictorial means.

However, I will argue without compromise that there is compelling support for the premises here at stake. B. C. van Fraassen (1991), for instance, clarified that the transfer of information is not necessarily bound to verbal means and measures, *even if* the debate *about* the transfer of information proceeds verbally for the most part. In other words, the partial restriction to verbal means does not conflict with the *possibility* of a non-verbal transfer of information:

> "That we cannot convey information, or say what a theory entails without using language, does not imply that – after all, we cannot say what anything is without language" (van Fraassen 1991, p. 5).

Hence, van Fraassen (1991) indicates that not all that is said (in the sense of *stated*) is necessarily said (in the sense of *expressed by virtue of words*), even if the equivocal use of the word 'saying' leads exactly to this (ultimately controversial) conclusion.

Further backing for the premises employed in this thesis comes from N. Goodman, whose remarks on analytical approaches complement the argument. Goodman argued that analytical approaches do by no means need to match *substantially* the objects of analysis (cf. Goodman & Lüdeking 2005, p. 263):

> "One cannot blame an analytical approach for not adapting to its object of analysis. That would be as pointless as if one would blame a description of grass for not being green"[42] (Goodman & Lüdeking 2005, p. 263).

Accordingly, given that any substantial equivalence between an analysis and its object is just contingent, one cannot infer from a verbal(-ised) approach the verbal 'nature' of the so-approached entities, such as non-verbal means.

On these grounds, I uphold the line of reasoning, which stresses that: *even if* reports or arguments on pictorial means are conveyed verbally or partially verbally, the intuition that there are genuine picture-like components or triggers, which co-constitute the means under consideration (resp. the contexts, where these means are taken from or where they are transferred into), remains untouched.

It remains to refine the strategies for representing picture-like means. In this respect, I suggest a combination of formal and interpretative constraints:

- In accordance with the source documents (i.e. Schittler et al. 2010; Falk et al. 2009), I will address the printed components under study as 'figures'. The only exception to this terminological alignment concerns the *tables* examined, which I added to the range of 'figures' in consideration of their figure-like separation of the context.

---

42 Translation of the following quote: „Man kann einer analytischen Untersuchung nicht vorwerfen, dass sie sich ihrem Gegenstand nicht anverwandelt. Das wäre ebenso unsinnig, wie wenn man einer Beschreibung von Gras vorwerfen wollte, dass sie nicht grün ist" (Goodman & Lüdeking 2005, p. 263).

- I furthermore decided to keep the inscriptions of the figures, in order to preserve at least the nearest parts of their original surroundings. I will try to diminish the confusion that arises from diverging numerations by means of graphical additions. For example, I added a black frame to the figures, in order to mark them (including their numerations) as encapsulated extracts.
- Regarding the interpretation: the main guideline here is that the extension of the term 'picture-like means' shall not be restricted to some square centimetres of dyed paper. Those quotable traits are called 'figures' (see above), but their status and functions as 'picture-like means' is debated below in respect of the in-depth analysis of their use, which has been approved by the paper's authors in terms of content and phrasing.
- Since the planned representation of representations does not work without surveying and matching the details about the contexts of use, I refrain from adopting any terms or phrases from the commenting literature that suggest a *non-pragmatic* view of picture-like means, associated, for instance, with terminological vehicles such as 'epistemic *thing*' ("*epistemisches Ding*", Rheinberger 2005, p. 66). The latter suggests that 'things' in the sense of physically describable objects can fully instantiate the notion of means. This, however, has been excluded above insofar as 'things' – without any addition pertaining to their projected use – are not subject to the discussion here.

3.1 Example 1: modelling mesenchymal stem cell differentiation

The first set of examples is taken from research on mesenchymal stem cells[43] conducted at the Institute for Systems Theory and Automatic Control at the University of Stuttgart. The source of information that I singled out for an in-depth analysis, along with the personal communication, is the paper "Cell differentiation modelled via a couple two-switch regulatory network" by D. Schittler, J. Hasenauer, F. Allgöwer & S. Waldherr (2010).[44]

---

43 Mesenchymal stem cells are progenitor cells which can differentiate into a range of cell types, including bone cells (osteoblasts) and cartilage cells (chondrocytes) cells (cf. Schittler et al. 2010, p. 1).

44 I wish to thank Daniella Schittler and her co-authors for the support and advice.

In order to approach the picture-like means involved in this document (resp. in *what* it documents), it is convenient to proceed step-by-step, starting with the analysis of the structure as defined by the subtitles; more detailed information will follow.

- The paper begins with the ***Introduction*** of the main points of departure; among them are the research objectives (cf. 3.1.1), the motivators for the overall research endeavour (cf. 3.1.2), and initial assumptions that are chosen either in accord with the state of research (cf. 3.1.3), or in deviation from previous approaches (cf. 3.1.4).
- The paper moves on with the introduction of a mathematical ***Model*** (cf. 3.1.5) that builds on the initial assumptions as announced in the preceding section.
- Having introduced the mathematical model, the paper proceeds with the ***Analysis*** of the model (resp. of its solutions). The analysis focuses on stability properties (cf. 3.1.6) and bifurcation properties (cf. 3.1.7).
- The paper continues with the analysis of ***Differentiation Scenarios***. The scenarios are specified for single cell switching (cf. 3.1.8) and cell population effects (cf. 3.1.9). The second scenario is based on a stochastic extension of the model.
- The paper concludes with the ***Discussion*** of the findings (cf. 3.1.10) and with further additions, such as the list of literature.

This overview is the preparation for a further degree of precision. Accordingly, I will add more details (in the proposed chronology) on each of the sub-titles and related sub-sections that I picked from the inspected paper.[45]

3.1.1 Research objectives

The overall research objective tackled in the paper under study (Schittler et al. 2010) is to explore the "functional mechanisms" (p. 1) that guide the "differentiation of progenitor cells [here: mesenchymal stem cells] into bone [osteogenic] or cartilage [chondrogenic] cells" (p. 1). The focus of research lies particularly

45 All quotes occurring in this section are from the studied paper (if not marked otherwise). On that account, only page numbers are added when it comes to signifying the quoted sources.

on the role of genes and genetic stimuli, which are suspected to determine the "fate" (p. 1) of the progenitor cells, that is, the cell type into which the progenitors differentiate.

The "key determinants" (p. 1) of the progenitor cell differentiation are explored by the aid of a mathematical model. The latter is intended to represent the hypothetical relation between (1) the behaviour of genes (or, more precisely, of "transcriptional regulators" [TRs] that determine genetic expressions), (2) stimuli supplies, and (3) the empirically warranted behaviour of differentiating progenitor cells (p. 1). The idea is thus to create a mathematical correlate for biological accounts of progenitor differentiation; one that fits empirical findings and adds further hypothetical conditions. *Whether* the model succeeds in this sense is tested with regard to the coherency and consistency of the factual assertions that are drawn on the one hand from the interpretation of the mathematical solutions, and on the other from the interpretation of the experimental findings.

If the acquired interpretations correspond with each other in this sense, it is said that the mathematical model "explains" (p. 2) the progenitor differentiation, because it allows for the subsuming of the biologically described cell behaviours into the mathematically described types of behaviours. In the philosophy of science, this kind of explanation can be classified as a "functional explanation" (according to Cummins 1975; cf. 2.2.1.2), for it is designed to illuminate the capacities of the to-be-explained process (here: mesenchymal stem cell differentiation) in consideration of the to-be-modelled and -explored functioning of the involved entities (here: of genes and genetic stimuli).

3.1.2 Motivators

Motivation is drawn from the prospect of controlled productions ("tissue engineering") and exploitations ("medical therapies") of organic material (p. 1). The special focus on mesenchymal stem cells – also known as "MSCs" (p. 1) – is justified with reference to the state of research: in the case of MSCs, "the key determinants guiding differentiation are still poorly known" (p. 1).

3.1.3 Initial assumptions (a)

In agreement with the state of research, the authors adopt a widespread measure of abstraction, which is needed since the real number of genes in a cell is hardly tractable by means of an analytical approach. The authors therefore suggest modelling the progenitor cell interior in accordance with "the widespread minimal motif of two genes [...] with mutual inhibition" (p. 1). In order to allow for such an inhibition between the genes as well as for auto-active or external (e.g. stimuli-based) impacts on this inhibition, a switch-mode is assumed. Accordingly, the inhibition is either active or inactive, while in the active state it can switch into the inactive state, and vice versa (until further limitation; cf. 3.1.5). Adopting this idea, progenitor cell differentiation is described as "a transition ["switch"] from one stable equilibrium point to another, each corresponding to a different cell type" (p. 1).

3.1.4 Initial assumptions (b)

In deviation from previous approaches, the authors employ a genetic "pre-switch" (p. 2) that changes the state of the cell in such a way that it becomes ready for a "fate-switch" (p. 2). This shall adjust the model to empirical results, especially to the finding that differentiation rates and outcomes correlate with the supply of stimuli, also *progenitor* stimuli. Furthermore, the authors take into account the observed "irreversibility" (p. 2) of the cell differentiation and suggest capturing it in terms of "activating and inhibiting factors" (p. 2). This is also said to be closer to the experimental results, compared to other accounts proposed so far (p. 2). The third innovative measure is devoted to the finding that cell differentiations do not evolve homogenously in cell populations. In order to obtain an adequate description of such "asymmetries" (p. 2) in cell behaviours, the model is thus further equipped for a stochastic simulation of differentiation scenarios.

3.1.5 Model

In order to come up with a mathematical description, the assumptions taken from biology are reformulated in a way that prepares for their transferral into mathematical conditions. Hence, it is set that "there are three cell types to be captured in terms of stable equilibria, namely a progenitor P, an osteogenic O and a chondrogenic C cell type. Each type is recognised by high levels of a characteristic TR, denoted as x_P, x_O, or x_C, respectively, but low levels of the other TRs" (p. 2). Accordingly, each cell type is set in correspondence to a specific genetic state, which in turn is correlated to numerically appropriable amounts or levels of state regulators (TRs). Ultimately there are three distinct sets of state variables.

At this point, the reformulation of the initial modelling assumptions is interrupted by a hint to Fig. 8. This figure is said to "depict" (p. 2) the above-listed conditions that are to be "captured" (p. 2) by the model. In fact, a closer look reveals that Fig. 8 might have had its own impact on the choice of biological conditions, as it allows for no more than three *definienda*, given that the graph offers merely three options for a high TR. On that note, Fig. 8 not only "depicts" (p. 2) the conditions in the sense of a one-to-one repetition, but it adds a subtle hint to the scope of graphically supported modelling options.

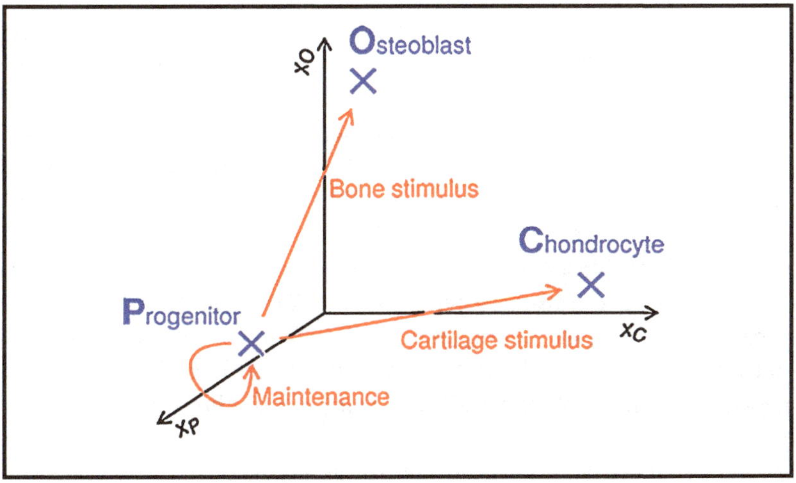

Fig. 8: Biological account of cell types and transitions (cf. Schittler et al. 2010, p. 2)

Close-up analysis of Fig. 8. Three black arrows spread out from a common point of contact, thus forming the three mutually orthogonal axes of a three-dimensional Cartesian coordinate system. Next to the arrowhead of each axis there is black lettering that individualises these axes as x_P, x_C and x_O. Each of the black axis pairs up with a blue cross and inscription. Axis x_P pairs up with a cross and the word '**P**rogenitor', axis x_C with a cross and the word '**C**hondrocyte' and axis x_O with a cross and the word '**O**steoblast'. Additionally, three red arrows point through the space that is unfolded by the axes. The red arrow next to '**P**rogenitor' is named 'Maintenance' and forms an open circle around the blue **X** nearby. The other red arrows are straight vectors. The steep one on the left side is called 'Bone stimulus' and points from '**P**rogenitor' to '**O**steoblast'. The other one on the right side is called '**C**artilage stimulus' and points from '**P**rogenitor' to '**C**hondrocyte'. Due to the listed elements, Fig. 8 is capable of accompanying the transfer of the biological conditions into mathematical conditions in the way that the figure effectively 'repeats' the conditions that need to be satisfied by the model. The key condition is that each cell type is defined as *distinct yet related* to the other types, which is realised in terms of a high value on one of the axes and low values on the other two, respectively. Additionally, red arrows define possible 'directions' of value changes, which can be construed as "state" (p. 1) or "fate" (p. 1) changes of the cell. Accordingly, Fig. 8 allows for recapitulating the conditions that are to be modelled; it even points subtly to the graphical rationale that might have justified the abstraction of *three* cell types, because, biologically speaking, MSCs could differentiate into more.

After having defined correlates for the three cell types (resp. for the related genetic states), the authors continue with the definition of the *relations* between them. For this purpose, they start with the conditions of irreversibility. These conditions must be involved since it "is observed in experiments that there occur certain transitions between the three cell types [whilst for others this is not the case]" (p. 2). It is, for example, possible for a progenitor *P* to turn into an osteoblast *O*, or into a chondrocyte *C*, or for it to change not at all. On the other hand, an osteoblast *O* cannot turn into a progenitor *P* ("dedifferentiation") or a chondrocyte *C* ("transdifferentiation") (p. 2). It is also not possible for a chondrocyte *C* to turn into a progenitor *P* or into an Osteoblast *O* (p. 2). These constraints are incorporated into the model in terms of "activating and inhibiting factors" (p. 2), which either allow for or prohibit the respective transitions of the genetic states.

Another set of assumptions is also important in this connection. As mentioned before, the authors aim to involve a genetic "pre-switch" (p. 2) that is sensitive to stimuli inputs. This is also achieved by the implementation of "activating and inhibiting factors" (p. 2): "an unspecific pro-differentiation stimulus will inhibit the progenitor maintenance factor, whereas stimuli acting in the pro-osteogenic or pro-chondrogenic direction enhance the activity level of the lineage-specific TR, respectively" (p. 3).

As suggested, the "activating and inhibiting factors" (p. 2) are used to establish relations between the components of the model. Fig. 9 is meant to "depict" (p. 3) them accordingly. Yet this time the figure does not allude to any graphically justified pre-configurations (cf. Fig. 8). Instead, Fig. 9 sticks to those components that are incorporated in the model.

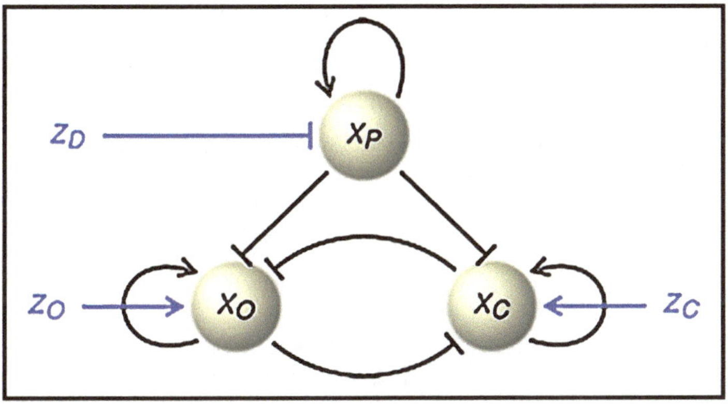

Fig. 9: Components incorporated into the mathe-matical model (cf. Schittler et al. C2010, p. 3)

Close-up analysis of Fig. 9. Three shiny grey balls are set up in a triangle. Black letters are pinned onto the 'belly' of each ball. The ball at the zenith is identified in this way as x_P, the ball on the left side at the bottom as x_O, and the ball on the right side at the bottom as x_C. The balls represent the equilibrium points (in mathematical terms) and the genetic states/cell types (in biological terms). The initial state variable x_P, which defines a progenitor, is situated at the zenith, the osteogenic state variable is situated on the left-hand side at the bottom, and the chondrogenic variable x_C is situated on the right-hand side at the bottom. Pointed black circular arrows are seen next to the balls and indicate that the gene activates its own gene expression (resp. state variable). Stump black arrows bridge between the balls and connect each ball with the other two. The stump arrowheads point to invariant inhibitions between those genetic states that cannot turn into each other, but that prohibit the options of "dedifferentiation" (p. 2) and "transdifferentiation" (p. 2). Blue arrows denote the stimuli for possible state switches. The stump blue arrow z_D signifies an inhibiting factor that decreases the progenitor's power to maintain its own state (resp. gene expression). Adding z_D accordingly makes a progenitor *ready* for switching. The sharp blue arrow z_O, on the other hand, denotes a stimulus that makes a ready progenitor turn into the osteogenic state. Analogously, the sharp blue arrow z_C signifies a stimulus that makes a ready progenitor turn into the chondrogenic state. Ultimately it appears that the relations between the model components can be repeated in view of Fig. 9. The latter accordingly tests well in listing and linking the instances and levels of description, namely the mathematical, genetic and the cell level.

For analogous reasons – at least in the sense of a summing-up tool – a "table" (p. 3; Fig. 10) is provided in order to present an exemplary set of parameters for the model. The distinct function of Fig. 10 is that it is the only explicit key to the parameters used. Their "meaning" (p. 3) and "values" (p. 3) are not made explicit in any other way in the paper.

TABLE I. Parameter set used for bifurcation analyses and simulations. The parameters are without units.

Parameter	Meaning	Value
n	Hill coefficient	2
a_P	Autoactivation of x_P	0.2
b_P	Basal activity of x_P	0.5
m_P	Inflection point for x_P	10
c_{PP}	Self-inhibition strength of x_P	0.1
k_P	Decay rate of x_P	0.1
a_O, a_C	Autoactivation of x_O, x_C	0.1
b_O, b_C	Basal activity of x_O, x_C	1
m_O, m_C	Inflection point for x_O, x_C	1
c_{OO}, c_{CC}	Self-inhibition strength of x_O, x_C	0.1
c_{OC}, c_{CO}	Mutual inhibition strength of x_O, x_C	0.1
c_{OP}, c_{CP}	Inhibition strength of x_P on x_O, x_C	0.5
k_O, k_C	Decay rates of x_O, x_C	0.1

Fig. 10: Exemplary parameter set (cf. Schittler et al. 2010, p. 3)

Close-up analysis of Fig. 10. Two horizontal double lines frame a nearly quadratic plane by adding a black ceiling and bottom to a three-columned list of black letters (left), black word-letter combinations (middle) and black numbers (right). No frame lines edge the left and the right side of the plane, leaving the table with open 'flanks'.[46] The same goes for the column in the middle, which is unframed as well. Between the ceiling's double line and a further horizontal line – hence in a separated row – there are inscriptions for the columns. The left column is called 'Parameter', the column in the middle is entitled 'Meaning', and the right column is called 'Value'. Altogether, the table provides a key – indeed the only one – to an exemplary parameter set for the mathematical model. The model is established (apart from the foregoing clarifications) by a set of differential equations (cf. Fig. 11). Fig. 10 and Fig. 11 are *not* addressed as

46 As announced in the foregoing paragraph 2.2, I added black frames to the figures in order to mark them (incl. their numerations) as encapsulated extracts. In this case, admittedly, the frame might distract the reader from the fact that the tablet is actually unframed on the left and right side.

'figures' in the paper. Still, it makes sense to add them in order to offer this (likewise spatially organised) information, which is central for guaranteeing the reproducibility of the results, as it presents the model in its computable rendition.

$$\frac{d}{dt}x_P = \frac{a_P x_P^n + b_P}{m_P + z_D + c_{PP} x_P^n} - k_P x_P, \tag{4}$$

$$\frac{d}{dt}x_O = \frac{a_O x_O^n + b_O + z_O}{m_O + c_{OO} x_O^n + c_{OC} x_C^n + c_{OP} x_P^n} - k_O x_O, \tag{5}$$

$$\frac{d}{dt}x_C = \frac{a_C x_C^n + b_C + z_C}{m_C + c_{CC} x_C^n + c_{CO} x_O^n + c_{CP} x_P^n} - k_C x_C, \tag{6}$$

Fig. 11: Model (cf. Schittler et al. 2010, p. 3)

3.1.6 Stability properties

The paper continues with the analysis of the mathematical model. As indicated before, the aim of the analysis "is to test whether the proposed model is able to reproduce the cell system properties observed during differentiation" (p. 3). In the first step, the analysis shows primarily that the model yields *exactly three stable solutions* (resp. "equilibrium points" or "equilibrium states"), and that it is thus capable of providing the required correlates for the three cell types P, O and C (p. 4). The corresponding proof is documented in the paper's appendix. Up to this point, it is just stated (not mathematically testified) that the model entails, indeed, "exactly three stable solutions" (p. 4).

The acquisition as well as the substantiation of this result depends *strongly* on figures. Fig. 12 is particularly intended to "illustrate" (p. 4) the corresponding numerical solutions of the model and thus to corroborate the claim that the model is "able to generate exactly three different stable equilibrium points" (p. 4). For this to be accomplished, Fig. 12 refers to Fig. 8 and Fig. 9 in the mode of specification (as it assigns functions [resp. their partial derivatives] to the model) and to Fig. 10 in the mode of application (as it presents a solution for this set of parameters). In the authors' terms: "Figure 3a [here: Fig. 12] illustrates the three-dimensional state space for an exemplary parameter set as in Table I [here: Fig. 10]. Comparing Fig. 3a [here: Fig. 12] against Fig. 1 [here: Fig. 8], one perceives the correct location of stable equilibrium points" (p. 4).

In anticipation of the discussion, it is to be noted that Fig. 8, Fig. 9, Fig. 10 and Fig. 12 apparently contribute to a whole network of mutual references. They also refer to external sources, such as the numerical solution of the model, which is only presented in the paper in figure-form. Personal communication also revealed that Fig. 12 is graphically idealised in order to have it better support the claim that the model allows exactly for three stable solutions. To this end, Fig. 12 presents only a reduced number of solutions, compared to its three-dimensional pre-version, which looks similar to Fig. 13.

In this context, it is also worthy of note that – due to complications with creating an apposite figure – the authors did not pursue the idea of a *four-dimensional* model, which would offer no intractable problems from the mathematical point of view. It thus turns out that figures can be essential to modelling decisions, even *normative* for them (cf. also Fig. 8).

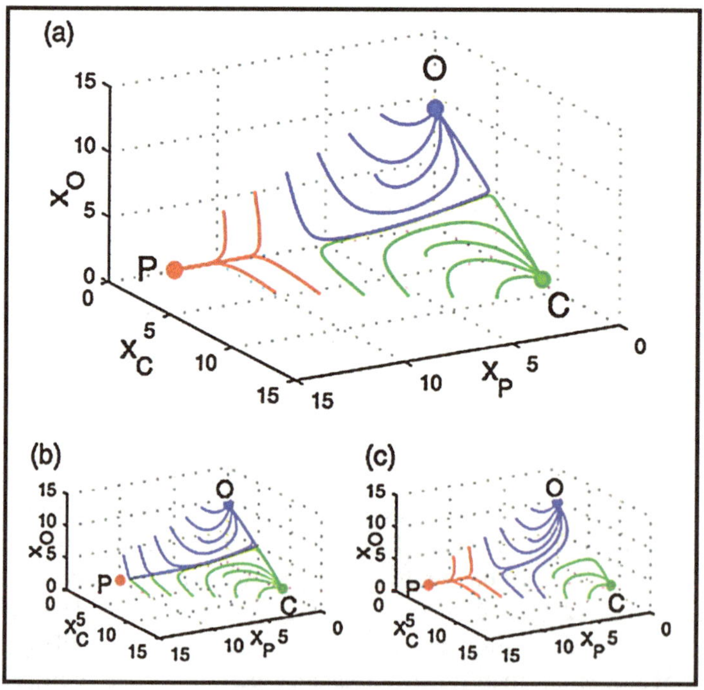

Fig. 12: Stable solutions (cf. Schittler et al. 2010, p. 4)

Close-up analysis of Fig. 12. This figure actually consists of a group of figures. One of the grouped figures is marked as (a). It sits above the other two figures (b) and (c), which it surpasses both in length and magnitude. In some other respects, the three (sub-)figures conform to one another. A number of blue, red and green lines traverse the three-dimensional space unfolded by the black axes of each figure, while each of the three coloured circles marks one of three stable solutions (i.e. "equilibrium points" or "states") (p. 4). The paper clarifies that Fig. 12 (esp. [a]) shall mainly support this claim, namely that the model has exactly three stable solutions, which can be used to represent the genetic states as defined by the biological model. Fig. 12 (esp. [b] and [c]) further supports the claim that the stable solutions change with the supply of stimuli. This correlation is drawn from and warranted by the variant sizes of the areas covered by lines in (a), (b) and (c). The variant lengths help substantiate, for instance, the claim that a high percentage of progenitor stimuli z_D reduces the area where the system is attracted to the stable solution that corresponds to the progenitor cell (marked with red lines), because under the impact of stimuli, the progenitors start differentiating. This is argued in particular with reference to figure (b), where the red lines are forced back completely by blue and green lines. A high amount of type-specific stimuli, on the other hand, increases the amount of the correlated cell types (e.g. osteoblasts; marked with blue lines, cf. esp. [c]). Hence, Fig. 12 takes part in the evaluation of the model. It is both an instance *and* warrant of the analysis, and as such it is used to derive and support assertions that help confirm that the model allows for the targeted mathematical conditions, which in turn are correlated to the biological ones.

3.1.7 Bifurcation properties

After having proved the existence of three stable solutions, the authors focus on the fact that the model also provides the requested correlatives for the *switches* of the genetic states (resp. for the differentiation of the associated cell types). Additionally, it captures effects of *stimuli* on these switches. Again, the related proof is provided in the appendix (p. 8f.); yet it is emphasised in advance that, "for the parameter set given in Table I [here: Fig. 10], the model allows exactly the required transitions" (p. 4).

It is in the body of the paper rather than in the appendix that two figures are mentioned that are used to establish the self-evidence of the claimed representational quality of the model. One of these figures has already been inspected (cf. close-up analysis of Fig. 12). The focus, here and in the paper, is therefore mainly on Fig. 14, which is meant to corroborate the claim that the model allows for *transition points* ("bifurcation points" in the mathematical language; p. 4), and for the fact that they vary due to stimuli inputs. The transition points can be found in Fig. 14 (a) (b); they are located where the differently shaped lines touch each other. These meeting points of the dashed lines and solid lines are hence set in correspondence with those situations in which a genetic state (resp. cell type) is about to switch (resp. to differentiate) (cf. close-up analysis of Fig. 14).

In addition to Fig. 12, which was used to confirm the existence of stable solutions and their sensitivity to stimuli inputs, Fig. 14 is accordingly offered for the sake of drawing an even stronger focus on the conditions of the genetic state switches.

It may thus happen – as in this case – that figures (here esp. Fig. 12 and Fig. 14) convey the same mathematical solutions, but pick out, or leave out, different information. The mathematical (resp. systems-biological) approach is thus strongly linked with the idea of generating different views of the model system. This is taken to be an advantage, since, precisely due to empirical accounts usually having time- and space-bound accesses on experimentally established systems, the maths-based approach stands out on account of the *variety of concurrent perspectives* drawn on the model system under consideration. This point was particularly stressed in the personal communication with the authors and elucidates their idea of a maths-based exploration of biological systems, which eventually relies heavily on the capacity of figures.

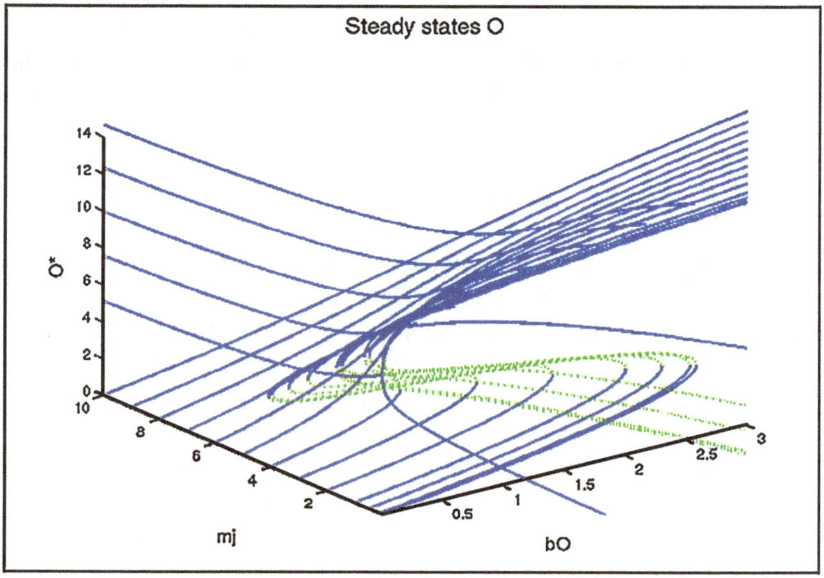

Fig. 13: Unpublished version (D. Schittler)

Close-up analysis of Fig. 13. Fig. 13 is one of the unpublished source figures of Fig. 14. Here, likewise, three black axes construct a three-dimensional space that is filled with coloured lines. The axes are individualised as **O*** (x_O), **mj** (z_D) and **bO** ($z_{O'}$). Axis **O*** touches axis **mj** and stretches from this point vertically to a value of 14. Axis **mj** bridges between axis **O*** and **bO** and reaches from the front where it touches **bO** to the area to the left of and behind the point at which it ties up with **O***. Axis **bO** shares a value point 0 with axis **mj** and reaches from there to a value of 3 on the right side in the background. The so-defined space is filled with traversing lines, of which most are solid and blue; only some (about five) turn partially into green dashed lines. In contrast to Fig. 14 and Fig. 12 the lines overlap with each other; two or three of them even cross axis **bO**. Personal communication revealed that the figure is actually implemented in 3D and that it can be rotated interactively, so that the overlaps change with the perspective. This also means that the published 2D version builds on an abstraction (resp. omission) of the original information. Without going further into these details, it stands out that it is difficult to use *precise* descriptors here. Instead, Fig. 13 suggests using vague phrases (containing "most of" or "about"). This hints at the reason why this figure was not deemed appropriate for publication: although one can surmise that there are equilibrium points (marked by the blue lines before and beyond the green lines) and points of transition (marked by the transition of blue lines into green lines), the figures' interior is too crowded for being scanned with the eyes. The authors thus decided to abstract markers for the equilibrium points and to project them into a plane. They deemed this to be sufficient as long as it supported the claim that there *are* equilibrium and transition points.

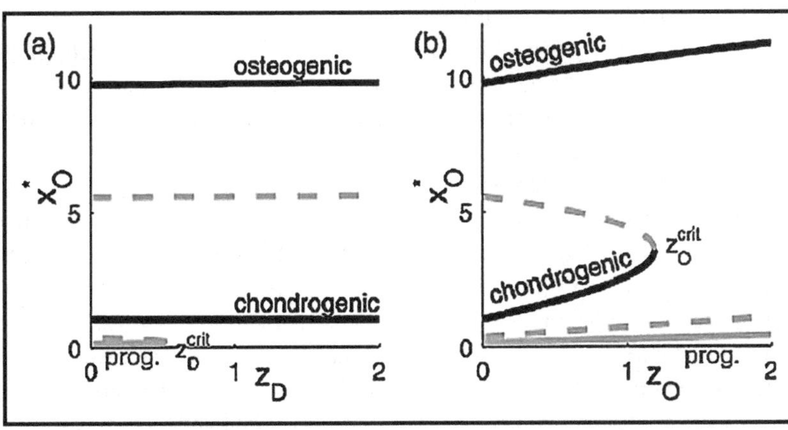

Fig. 14: Points of transition (cf. Schittler et al. 2010, p. 4)

Close-up analysis of Fig. 14. The figures that compose Fig. 14 are both referred to as "bifurcation diagrams" (p. 4). As such, they are expected to show that and how the related set of functions reacts to changes of parameters with changes in 'behaviour', that is, with qualitative changes of the solutions. In fact, Fig. 12 could be assigned to the same type. Yet, the label is reserved for this two-dimensional version in the context of the paper. In Fig. 14 (a), to begin with the diagram on the left, two solid black lines rest in a horizontal position within the plane, which is generated by the horizontal axis z_D and the vertical axis x_O. A short grey line – referred to as 'prog.' – reaches out from a point 0 and stops near 1, whereupon it 'turns back' and takes on the form of a dashed line. The point where the line 'turns back' is the point of transition, that is, the point where the system switches in qualitative respects. In biological terms this means that concentrations of progenitor stimuli higher than z_D^{crit} cause a progenitor cell to differentiate. Fig. 14 (b) indicates analogously that percentages of osteogenic stimuli z_O higher than z_O^{crit} cause the switch into an osteogenic cell. And since it is also assumed that chondrogenic stimuli have analogous impacts, the two figures (a) and (b) are considered sufficient to substantiate the claim at stake, namely that the model offers indeed correlatives for the impact of stimuli on the switching behaviour. As in Fig. 12 this result is partially guaranteed by *external* sources (here: model solutions), in contrast, for example, to Fig. 9, which refers entirely to notes and sources that are presented in the paper.

3.1.8 Single cell switching

The last parts of the paper are devoted to demonstrating "how [the] model can also serve for more descriptive applications to the biological problem" (p. 5). This means that the model is also tested as to whether it can "reproduce" (p. 3) "asymmetries" (p. 5) that occur in empirical settings. There, the term 'asymmetry' is applied to ill-explained variations.

The authors start by modelling the asymmetric behaviour of *single* progenitor cells. They find that two factors can be correlated with asymmetric behaviours: (1) variations of stimuli supplies, for instance in terms of delays, different orders or combinations; and (2) an "intrinsic bias" (p. 5) that makes the progenitor cell favour a lineage. Both factors are incorporated into the model, using varied parameter values.

Adding these factors yields that "[n]one of the two inputs z_D [progenitor stimulus], z_O [osteoblast stimulus] alone is sufficient to execute a transition guided toward one specific cell type *O* or *C*" (p. 6), since, for example, a high percentage of progenitor stimuli makes progenitors differentiate into osteoblasts or chondrocytes *before* lineage-specific stimuli can actually unfold an impact on the "fate" (p. 6) of the system (cf. Fig. 15 [b]). Hence, the reproduction of asymmetries gives justification to the suspicion that "in order to steer the system on purpose, a combination of both inputs is necessary" (p. 6).

Again, a *set* of figures is provided in this context (cf. Fig. 15). Like Fig. 12 and Fig. 14, Fig. 15 "depict[s]" (p. 5) numerical solutions for adjusted parameter sets. The difference is that these solutions are transferred *retrospectively* – by way of a test-run interpretation – from the mathematical into the biological language. Accordingly, on the basis of the afore-confirmed representational quality of the model, the model is also used to generate (resp. extract) mathematical implications, which are tested as to whether they allow for *further* hypotheses on the biological system. In accordance with this, the authors speak of "observations" (p. 5) when realising features that are as yet unaccounted for in the biological reference contexts. Given this, Fig. 15 is obviously treated as a *primary source* of the exploration of the model – that is, of the acquisition of further hypothesis on the modelled system. This stands in contrast to cases where figures are used to 'repeat' or 'depict' what was conceptualised before (cf. e.g. Fig. 9).

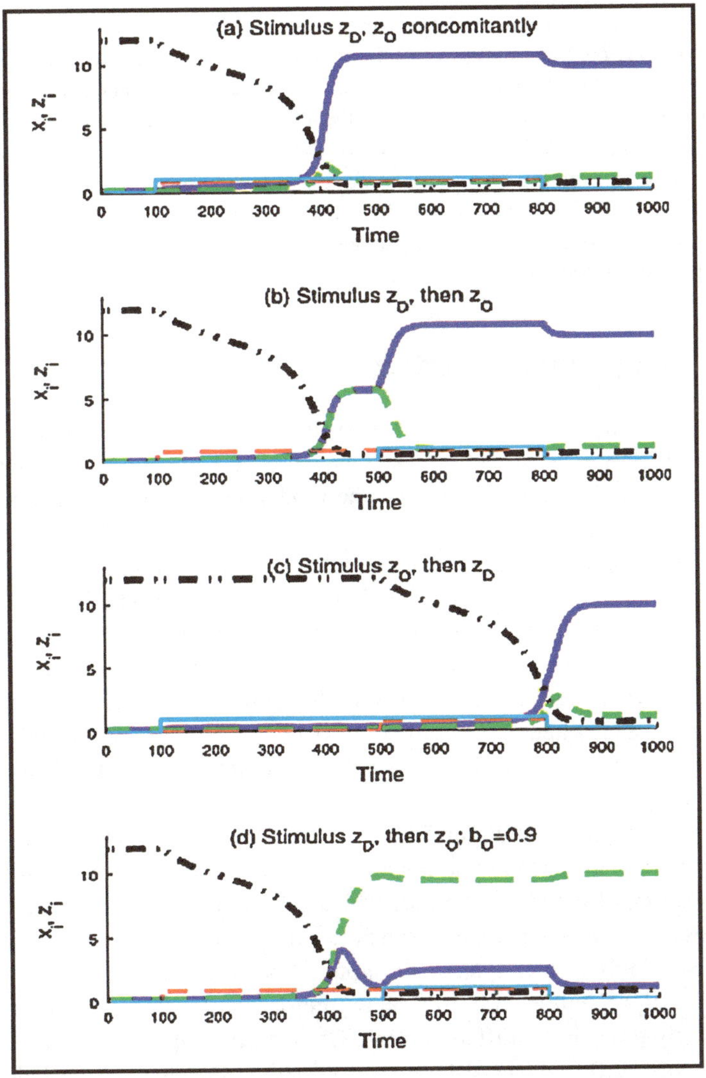

Fig. 15: Scenarios for different stimuli inputs (cf. Schittler et al. 2010, p. 5)

Close-up analysis of Fig. 15. Here again, a *set* of figures is compiled. This set consists of four figures that are in accordance with each other in terms of the basic architecture: solid and dashed lines between two axes. In this respect, Fig. 15 equals other sets of figures (esp. Fig. 14). A *distinct* feature of Fig. 15 is the variety of colours present, which allows for the discerning of a larger quantity of different lines, compared to Fig. 14. A further difference is that Fig. 15 lines up the figures in a vertical row. This arrangement makes sense in several respects: first of all, scenario analyses require more than one scenario, in order to offer a solid base for the comparison; yet only three figures fit into a horizontal row here (cf. Fig. 16), which might explain why the authors opted for the vertical row. On the other hand, the latter also allows for projecting vertical lines that cut through the figures and that help to compare the solutions with regard to the time values noted on the horizontal axes. In terms of this analytical support, Fig. 15 proves essential for the approach, for it aids the extraction of mathematical implications and guides their transference into biological hypotheses. On this basis, it is argued, for instance, that temporal factors, such as variant orders of stimuli inputs, can have an impact on a progenitor cell's "fate" (p. 1).

3.1.9 Cell population effects

The previous section focused on "extrinsic signals" (p. 5), especially on *stimuli inputs*, which are suspected to cause asymmetric behaviours. Additionally, it is now argued in the final section that the model also allows for "intrinsic" (p. 6) "fluctuations" (p. 6).

The latter occur in cell populations where the differentiation process may occasionally or partially not comply with the predictions derived from the modelled "mechanisms" (p. 1). This seemingly random behaviour is therefore of particular interest to researchers, as it "can impose limitations on the differentiation success rate" (p. 6). However, modelling seemingly random behaviour is also particularly challenging, as it eludes deterministic accounts. Consequently, the authors modify the model so as to extend it to "a set of stochastic differential equations" (p. 7; cf. Fig. 17).

As expected, the "simulation" (p. 7) is in line with the modelling assumptions. In other words, it turns out that the model is indeed capable of generating "noise" (p. 6), that is, of random solutions, which can be used to represent ill-explained behaviours as observed in experiments. The inspection of the 'random' solutions also yields further systematic inputs. For example, the authors find that a level of z_D [progenitor stimuli] "below the deterministic threshold" (p. 7) does not entirely prevent the differentiation. On that score, it is assumed that progenitor cells also differentiate at their own initiative, that is, without being in need of progenitor stimuli. The analysis furthermore suggests that a high value of z_D

[i.e. progenitor stimuli] can induce the differentiation even before "fate" (p. 3) stimuli unfold their directing impacts on the cell fate. Hence, the simulation indicates that the so-called 'asymmetric' cell population effects do not only depend on stimuli, but also on intrinsic factors that are only reproducible with the aid of stochastic measures at the recent stage of research.

Fig. 16 is said to allow for "seeing" (p. 7) exactly what is claimed, namely "that the percentages of cells depend on the amounts of stimuli z_D and z_O in a highly nonlinear way" (p. 7). Still, the figures also substantiate the assertion that – according to the model – there is at least a strong correlation between the amount of progenitor stimuli z_D and the amount of progenitor cells, whereas the amounts of bone and cartilage cells depend on the very combination of all types of stimuli (cf. close-up-analysis of Fig. 16).

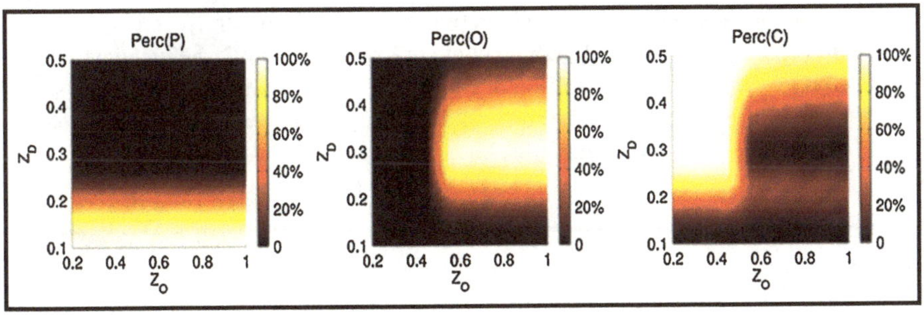

Fig. 16: Exemplary solution of the stochastic simulation (cf. Schittler et al. 2010, p. 7)

Close-up analysis of Fig. 16. Fig. 16 lines up three figures in a horizontal row. As in the sets before, the figures conform to each other in terms of their basic design, but differ from each other in terms of that which is portrayed between the axes. The part between the axes (z_D and z_O) forms a flat plane, which is in this case completely shaded with the colours black, red, yellow and white. Accordingly, no distinct elements (e.g. in terms of foreground and background parts) are marked out; even the 'white' area instantiates the colour code on the right. The transitions between the colours are smooth and partially blurred: black dots, for instance, appear in the red parts and, vice versa, red dots appear in the black parts, giving the 'melted' areas a shaded, indefinite colour. The scale next to the right flank of each figure assigns the colour *black* to low percentages of the cell type under consideration, *red* and *yellow* to higher percentages and *white* to the highest percentages, while each figure concentrates on a different *type*. Accordingly, the figure on the left 'Perc(P)' focuses on progenitors, the figure in the middle 'Perc(O)' focuses on osteoblasts and the figure on the right side 'Perc(C)' on chondrocytes. The significance of figure 'Perc(P)' gives rise to the claim that the amount of progenitor cells depends on the amount of progenitor stimuli, yet *not* on the amount of osteogenic stimuli, since the black area which 'rises up' from a value point higher than 0,2% on the vertical axis, shows no variances on the horizontal axis. The figure in the middle, on the other hand, is linked with the assertion that the amount of osteogenic cells depends on both – a sufficient amount of progenitor stimuli (about 0,2% on the vertical axis) and a sufficient amount of osteogenic stimuli (more than 0,6% on the horizontal axis). However, the highest percentages of the two stimuli do not yield the highest number of osteoblasts. The interpretation of this finding is that a high percentage of progenitor stimuli induces the differentiation (giving rise to osteogenic *and* cartilage cells), *before* the supplied osteogenic stimuli unleash their effects. This at least explains why the white part lumps in the middle, seen from the vertical axis. Finally, figure 'Perc(C)' is related to the claim that the number of cartilage cells correlates with a high amount of progenitor stimuli (vertical axis) and a low amount of the rivalling osteogenic stimuli (horizontal axis). The percentage of cartilage cells is thus highest on the upper left-hand side, which indicates high progenitor stimuli and low osteogenic stimuli. In other respects, the behaviour of the cartilage cells is treated as analogous to that of osteogenic cells within (most parts of) the paper. Hence, when adding high amounts of progenitor and high amounts of cartilage stimuli, the outcome is to be considered as equivalent to that of figure Perc(O). This treatment holds analogously for the last figure of the paper, which shows an exemplary result for additional time parameters (cf. close-up analysis of Fig. 18).

$$dx_P = -k_P x_P dt + \frac{a_P x_P^n + b_P}{m_P + z_D + c_{PP} x_P^n}(dt + \sigma dW_P), \tag{11}$$

$$dx_O = -k_O x_O dt + \frac{a_O x_O^n + b_O + z_O}{m_O + c_{OO} x_O^n + c_{OC} x_C^n + c_{OP} x_P^n}(dt + \sigma dW_O), \tag{12}$$

$$dx_C = -k_C x_C dt + \frac{a_C x_C^n + b_C + z_C}{m_C + c_{CC} x_C^n + c_{CO} x_O^n + c_{CP} x_P^n}(dt + \sigma dW_C), \tag{13}$$

Fig. 17: Stochastic simulation: model (cf. Schittler et al. 2010, p. 7)

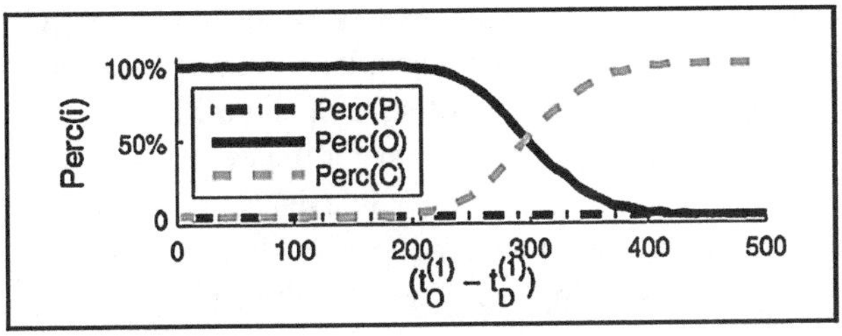

Fig. 18: Simulation for additional time parameter (cf. Schittler et al. 2010, p. 7)

Close-up analysis of Fig. 18. "After comparing the amounts of stimuli, [the authors] also considered the effect of their timing" (p. 7). Fig. 18 is said to "depict" (p. 7) the results of the simulation accordingly for an exemplary set of time parameters. With this, it serves (like Fig. 15 and Fig. 16) as a basis for the derivation of implications pertaining to the biological system, this time concerning the impact of stimuli delays. In terms of the composition, Fig. 18 apparently takes over some of the elements of the other figures, but combines them in a unique way, starting with a completely black and white colour scheme. Moreover, the numbers of lineage-specific stimuli are specified on the vertical axes and related to the time values on the horizontal axes. The interior between the axes is filled with three differently shaped lines (represented as black solid, black dashed, grey dashed) that reach from time value 0 on the left to value point 500 on the right, each 'following' a different curve-like route. The lines indicate the impact of stimuli inputs on the different cell types over time, as can be derived from the quadratic key to the figure, which is pressed between the curves. The interpretation tells us that a delayed input of progenitor stimuli z_D can be treated as analogous to an increase of progenitor stimuli (cf. Fig. 15), whereas the osteogenic stimuli are only effective if the progenitor stimuli are not delayed. In other words, if a high amount of osteoblasts is being targeted at, progenitor stimuli must be supplied first and osteogenic stimuli quickly thereafter, in order to guide the system in the appointed direction. And so, to finish the inspection of the figures with a last remark: Fig. 16 and Fig. 17 may both be considered 'multifunctional' in the sense that they serve as a testimony of the simulation (resp. its solution) and also as a basis for the extraction of mathematical implications, which are then transformed into hypothesis and operational recommendations for future biological experiments (cf. 3.1.10).

3.1.10 Discussion

The authors close with a positive account of the approach, as it worked out "to reproduce [...] three stable equilibrium states (cell types) and the two transitions when the cell leaves the progenitor state and adopts osteogenic (bone) or chondrogenic (cartilage) lineage" (p. 8). More abstractly, the alignment of the mathematical and the biological conceptions of the progenitor cell differentiation was found to be possible. Moreover, an explanation was found for "why MSC differentiation requires a combination of several stimuli [...], and why the amount and sequence of stimuli might be relevant for the eventual differentiation success rate" (p. 8). On this basis, the authors could even bring up some advice for future experiments. Above all, they recommend adjusting experimental protocols to a more accurate "stimulus timing" (p. 8) and "combination" (p. 8).

3.2 Example 2: modelling signal transduction in mammal cells

The second set of examples is taken from research on interactive visualisation techniques conducted at the Institute of Visualisation and Interactive Systems at the University of Stuttgart. The source of information, which I chose in this case – along with the personal communication – is the paper entitled "Visualization of Signal Transduction Processes in the Crowded Environment of the Cell" (2009) by M. Falk, M. Klann, M. Reuss and Th. Ertl.[47] This paper is a suitable follow-up to the one presented before (cf. 3.1), because it employs further types of figures, some of which have not been considered in literature so far, such as "microscope-like images"[48] (p. 169). A further advantage is that the authors comment explicitly on the benefits of at least some of the figures. Figures thus serve as means *and* issues here, which makes the paper ever more valuable for the tasks pursued in this thesis. Finally, another advantage is that the paper ties in smoothly with another approach from the domain of systems-biology. This facilitates the resumption of the analysis due to some analogies in the inspected cases of research. But before coming to the in-depth analysis, I will again offer a brief overview of the paper's structure as defined by the subtitles (as done in chapter 3.1).[49]

- The paper commences with an ***Abstract*** and the ***Introduction***, which introduce the main objectives and motivators of the approach (cf. 3.2.1).
- The paper continues with a brief overview of the methods that are either adopted from ***Related Work*** or suggested in response to it (cf. 3.2.2).

47 I wish to thank Martin Falk and his co-authors for the support and advice.

48 In fact, the idea is not to mimic microscopes, but microscopic images.

49 All quotations occurring in this section are from the studied paper (if not marked otherwise). Therefore, only page numbers are added when it comes to signifying the quoted sources (as in the chapter before).

- The section ***Visualization and Simulation as Tools in Systems Biology*** begins with general remarks on the combination of simulation and visualisation techniques (cf. 3.2.3). The following sections then add more detailed information on the rendering strategies that are employed in order to represent the (otherwise biologically described) target system. The first section ***Schematic Illustration*** introduces general codes (cf. 3.2.4). The next sections add corresponding information on the ***Proteins and Trajectories*** (cf. 3.2.5), ***Reactions*** (cf. 3.2.6), ***Cuts and Transections*** (cf. 3.2.7), ***Depth of Field and Depth Cues*** (cf. 3.2.8) and ***Microscopic Images*** (cf. 3.2.9).
- The section ***Application*** reports on two application cases, namely the representation of different transduction mechanisms for ***Mitogen-Activated Protein Kinase*** (cf. 3.2.10) and the ***Delivery of Drug Molecules in Solid Tumours*** (cf. 3.2.11).
- The fifth section reveals background information on ***Technical Aspects*** pertaining to the implementation of the simulation, the visualisation framework and the interaction (cf. 3.2.12).
- The conditions (esp. data acquisition, input and parameters) under which the approach has proven feasible and efficient are exposed in terms of ***Results*** (cf. 3.2.13).
- The paper ends with notes on the ***Conclusion and Further Work*** and with standard additions, such as the list of literature (cf. 3.2.14).

The sections listed will now be fleshed out with further details, in order to approach the figures that are involved in them (resp. in the documented research).

3.2.1 Research objectives and motivators

The overall objective of the research project under consideration (Falk et al. 2009) is to "model and analyze cellular signal transduction" (p. 169) in such a way that it becomes attuned to recent biological research and that it facilitates the search for further hypotheses about the target system. For this purpose, the targeted "cellular events" (p. 169) (resp. the biological accounts thereof) are represented by means of a stochastic simulation and "advanced visualization techniques" (p. 169).

The authors are motivated by the desire for a better understanding of "how signals reach their targets" (p. 169). This is of potential value to medical research on controlled drug delivery. Further expectations are raised in the realms of scientific visualisation where researchers seek to prove that the "systems biology approach can only lead to a better understanding of the biological processes if we are able to visualize the results and relate it to the images and outcomes of experiments" (p. 169).

Exemplary 'show-pieces' of the visualisation-focused approach are lined up on top of the introductory parts of the paper, where they allude – almost motto-like – to the integrated methods and related areas of expertise, namely the experimental, systems-biological and the computer-graphics field (cf. close-up analysis of Fig. 19).

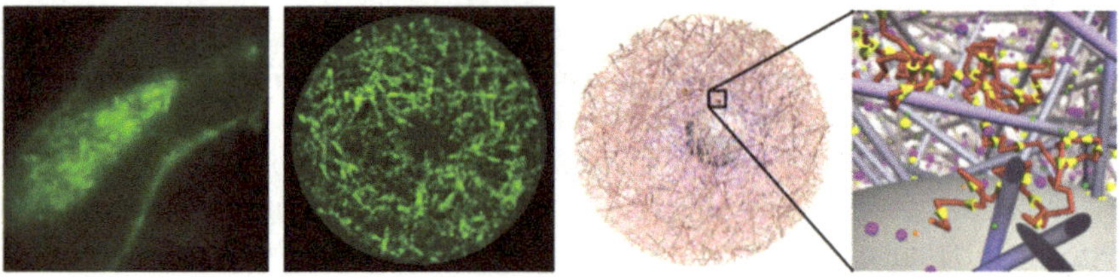

Fig. 19: Introductory sampling of screenshots (cf. Falk et al. 2009, p. 169)

Close-up analysis of Fig. 19. The first figure (resp. sampling of figures) is placed right above the introductory parts of the paper, indicating the importance of figures in the documented approach (Falk et al. 2009). The context reveals that Fig. 19 lines up four figures that were extracted from the modelling procedures presented in the paper. Accordingly, Fig. 19 entails different extractions or testimonies of the modelling endeavour presented in the paper. The figure on the very left is introduced as a "microscopic image from an experiment obtained with confocal laser scanning microscopy" (p. 169). This image has been given epistemic priority in the sense that the others on its right are approved with reference to it. This might also be the reason why it occurs first in the direction of reading). The second figure (centre left) is referred to as a "microscope-like image" (p. 169). This description indicates that it has been created on the basis of simulation results but graphically adjusted to the one on its left (cf. 3.2.9). By contrast, the figure on its right is introduced as a "geometric representation" (p. 169). The following sections (cf. 3.2.5ff.) reveal that the latter is meant to illustrate the possibility of exploring the model (resp. its visualised solutions) by studying the glyphs that represent the "tens or hundreds of thousands [of modelled] objects" (p. 170). Finally, the last figure on the right is meant to illustrate the possibility of *zooming into* or *out of* the glyph-based rendition. For this purpose, it is linked to the figure on its left by means of a black frame and two lines, which symbolise the acts of zooming in (from the left to the right) and zooming out (from the right to the left). Further information on the figures is provided in the body of the paper (see further below).

3.2.2 Related work

The project builds upon previous work, especially in terms of the simulation, but overall it is intended to present a new combination of measures. This reportedly includes the introduction of interactive 3D animation into bio-chemical network modelling, where it has formerly been avoided due to the high "visual complexity" (p. 170) and due to the related problems of tractability. The complexity results from the high number of components of the model, such as objects, relations and events (cf. p. 170). The authors suggest fostering the assimilation of these large amounts of complexly related information by means of different rendering strategies, such as "segmentation, depth cues and depth of field" (p. 170; cf. 3.2.3ff.). By these means, the authors attempt to overcome a notorious deficit concerning the use of graphs. Usually, graphs are deemed convenient for plotting numerically specified information, such as the "number of activated molecules at a given time" (p. 170). However, the authors object that graphs do not aid the representation of spatial information, such as about the *locations* and *individual pathways* of the model components (here: molecules). Therefore, the aims are to take care of *every* component of the modelled cell, instead of just an average number of them, and to foster the analysis of the model by providing access to the spatial information about the model components.

3.2.3 Techniques combined

The authors make use of different methods. Their choice is an agent-based Monte-Carlo simulation that combines a focus on *single objects* with the virtues of a *stochastic* approach, that is, with the capacity to represent *asymmetric* system behaviour, such as irregular particle distribution. Moreover, in order to represent both temporal *and* spatial aspects of the targeted system, the simulation builds on partial differential equations, which allow for the required number of variables.

On this basis, the simulation is supposed to fit the biological descriptions of the target system as gained from experimental accounts with "fluorescence microscopy" (p. 170; cf. 3.2.9). Accordingly, the idea is that empirical approaches settle the terms of modelling so that, "[i]deally, the outcome of the agent-based Monte Carlo simulation […] should give the same images when the agents are visualized. The agents should behave like the molecules in a cell" (p. 170).

3.2.4 Schematic illustration of the cell model

The graphical rendition of the computed solutions is complicated by the need to balance the mathematical complexity with the complexity that researchers are able to cope with when they examine the solutions on a visually aided basis. In response to this challenge, the authors propose *simplifications* in both domains.

Firstly, "[t]he highly sophisticated nature of a cell is simplified in the following way: it consists only of proteins, obstacles, and a generalized cytoskeleton. [...] [The] proteins, peptides, and hormones, all part of the signal transduction process in cells, are mapped to spheres, which are large enough to enclose their [i.e. the protein's, peptide's and hormone's] structure. Furthermore, we have spherical obstacles to simulate other subcellular components [...]. The filaments of the cytoskeleton are represented by elongated cylinders, which are placed randomly in the cell. During the simulation the cytoskeleton is stationary" (p. 170).

In addition to the simplification of the cell interior, "[t]he proposed visualization technique employs geometric objects and glyphs to display these simplified cell components" (p. 170). This means that each biological *type* of entity is set in correspondence to a geometrical *type*, which implies that individual traits are omitted in order to reduce complexity. The resulting three-dimensional map of *typified* cell components is referred to as a "schematic illustration" (p. 170) of the simulation (resp. its solutions).

3.2.5 Rendering scheme for proteins and trajectories

Among the modelled components, *proteins* are of particular interest, for it is assumed that proteins have the function of signals. The authors therefore aim to account for a variety of protein behaviours, including their ways of moving through the cell (i.e. by diffusion or by "active motor transport", p. 170; cf. 3.2.10) and their interaction with other cell components.

In order to facilitate the assimilation of the computed solutions, researchers are involved in the rendering process, because experience has shown that the solutions (resp. approximations) of the underlying mathematical model are not efficiently processed when they are generated automatically and presented as matrices or rows of numbers (cf. Fig. 20).

In the case under consideration, the high visual complexity even calls for *further* aids in addition to the transcription of the solutions into a geometrical, topological or graphical coding scheme; since, for instance, "[s]ome proteins are too small to be clearly visible at the desired resolution of the cell" (p. 170). For this reason, there is the additional option to "scale[...] up" (p. 170) a selected set of proteins. Moreover, the events of interest (e.g. the movements of signals) are marked with particular glyphs, which foster the recognition (e.g. "thin illuminated cylinders") (p. 170). This is also why "small arrow tips" (p. 170) are chosen in order to "highlight the direction of proteins at particular segments" (p. 170). Besides, in order to facilitate the focus on selected entities, further rendering measures are employed (cf. close-up analysis of Fig. 21). Researchers are also able to define the intervals of the animation, which is crucial for the tractability of the rendition, because the "simulation [...] consists of more than 1 million time steps" (p. 170).

In order to gain an idea of the resulting rendition, two figures are offered as demonstrators (Fig. 21 and Fig. 22). Each of them provides screen-shots of the "virtual cell" (p. 169). However, the obvious problem about Fig. 21 and Fig. 22 is that these figures cannot expose all features and facilities described in the paper, because they are fixed, two-dimensional *extractions*, which is to say that they suffer from a lack of information compared to the interactive, three-dimensional source figures. (In fact, I only know of *one* publication that tries to compensate for this drawback by adding a compact disc with further digital material [i.e. Beyer & Lohoff 2005]. Others, for instance, provide links to demonstrators on a webpage).

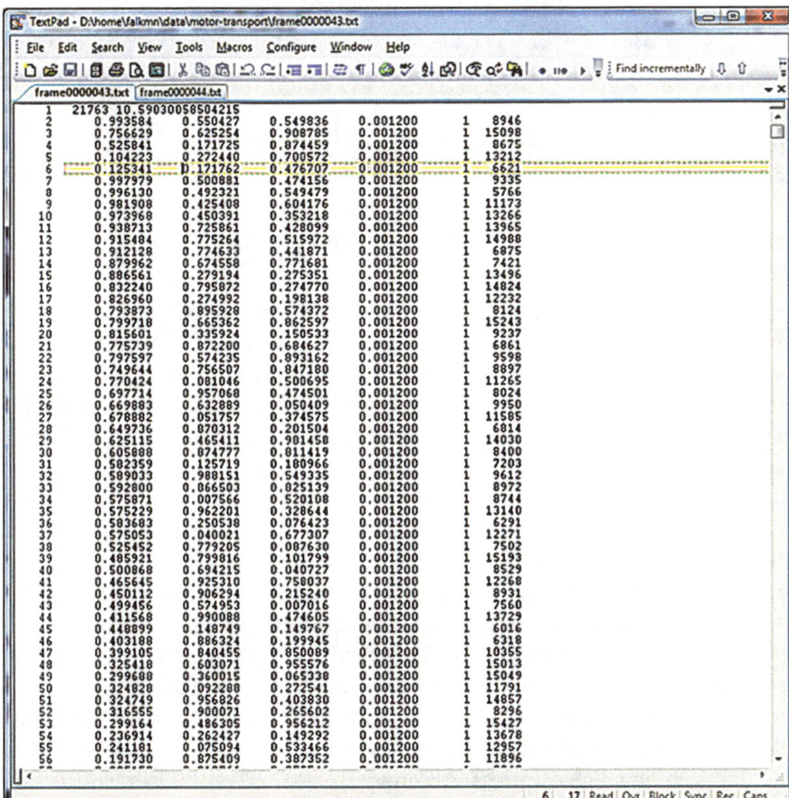


```
21763 10.59030058504215
0.993584  0.550427  0.549836  0.001200  1   8946
0.756629  0.625254  0.908785  0.001200  1  15098
0.525841  0.171725  0.874459  0.001200  1   8675
0.104223  0.272440  0.700572  0.001200  1  13212
0.125341  0.171762  0.476707  0.001200  1   6621
0.997979  0.500881  0.474156  0.001200  1   9335
0.996130  0.492321  0.549479  0.001200  1   5766
0.981908  0.425408  0.604176  0.001200  1  11173
0.973968  0.450391  0.353218  0.001200  1  13266
0.938713  0.725861  0.428099  0.001200  1  13965
0.915484  0.775264  0.515972  0.001200  1  14988
0.912128  0.774633  0.441871  0.001200  1   6875
0.879962  0.674558  0.771681  0.001200  1   7421
0.886561  0.279194  0.275351  0.001200  1  13496
0.832240  0.795872  0.274770  0.001200  1  14824
0.826960  0.274992  0.198138  0.001200  1  12232
0.793873  0.895928  0.574372  0.001200  1   8124
0.799718  0.665362  0.862597  0.001200  1  15243
0.815601  0.335924  0.150533  0.001200  1   9237
0.775739  0.872200  0.684627  0.001200  1   6861
0.797597  0.574235  0.893162  0.001200  1   9598
0.749644  0.756507  0.847180  0.001200  1   8897
0.770424  0.081046  0.500695  0.001200  1  11265
0.697714  0.957068  0.474501  0.001200  1   8024
0.669883  0.632889  0.050409  0.001200  1   9950
0.678882  0.051757  0.374575  0.001200  1  11585
0.649736  0.870312  0.201504  0.001200  1   6814
0.625115  0.465411  0.901458  0.001200  1  14030
0.605888  0.874777  0.811419  0.001200  1   8400
0.582359  0.125719  0.180966  0.001200  1   7203
0.589033  0.988151  0.549335  0.001200  1   9612
0.592800  0.066503  0.825139  0.001200  1   8972
0.575871  0.007566  0.520108  0.001200  1   8744
0.575229  0.962201  0.328644  0.001200  1  13140
0.583683  0.250538  0.076423  0.001200  1   6291
0.575053  0.040021  0.677307  0.001200  1  12271
0.525452  0.779205  0.087630  0.001200  1   7502
0.485921  0.799816  0.101799  0.001200  1  15193
0.500868  0.694215  0.040727  0.001200  1   8529
0.465645  0.925310  0.758037  0.001200  1  12268
0.450112  0.906294  0.215240  0.001200  1   8931
0.499456  0.574953  0.007016  0.001200  1   7560
0.411568  0.990088  0.474605  0.001200  1  13729
0.448899  0.148749  0.149767  0.001200  1   6016
0.403188  0.886324  0.199945  0.001200  1   6318
0.399105  0.840455  0.850089  0.001200  1  10355
0.325418  0.603071  0.955576  0.001200  1  15013
0.299688  0.360015  0.065338  0.001200  1  15049
0.324828  0.092288  0.272541  0.001200  1  11791
0.324749  0.956826  0.403830  0.001200  1  14857
0.316555  0.900071  0.265602  0.001200  1   8296
0.299164  0.486305  0.956212  0.001200  1  15427
0.236914  0.262427  0.149292  0.001200  1  13678
0.241181  0.075094  0.533466  0.001200  1  12957
0.191730  0.875409  0.387352  0.001200  1  11896
```

Fig. 20: Unpublished source file (M. Falk)

Close-up analysis of Fig. 20. This screenshot is taken from a *non-published* source file. It presents a tiny selection of the computed solutions in such a way that it leaves room for improvements of the visual rendition; because, experience has shown that even a trained person struggles with the interpretation of such matrices, even more so as this is just *one* screen-shot of a 4,3 MB file, which in turn corresponds to only *one* time step of the simulation, which in turn is still small compared to high-performance simulations. Thus, even if one can acquire some ideas about the target system's behaviour from this matrix, it seems hopeless to defend them on this basis. The demands on data assimilation call rather for "advanced visualization methods" (p. 169), not to make the assimilation more efficient but simply to make it possible *at all*. As argued below (cf. 4.3), the success of visualisation in the area of data rendition is due to its specific potential to render the *relations* between data points in a more convenient way (cf. 4.3.3.1; also Krempel 2005).

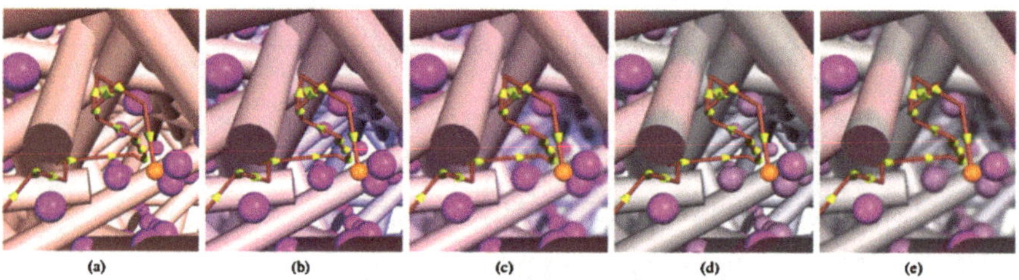

Fig. 21: Variations of graphical rendition (cf. Falk et al. 2009, p. 171)

Close-up analysis of Fig. 21. Fig. 21 lines up *five* variations of *one* particular screenshot, indexed as "[a]", "[b]", "[c]", "[d]" and "[e]" respectively. Each of the screenshots presents the same data, and it is interpreted as a representation of parts of a protein's trajectory. The close-up inspection reveals that the screenshots differ in graphical terms, although they are congruent in terms of those characteristics that can be described with reference to geometrical and topological categories. Accordingly, the analysis of the first screenshot "[a]" (counted from the left) suffices to capture *some* of the common aspects. These include a little orange ball which occurs only once in each screenshot and which is to represent the signalling molecule. Moreover, the common features also include a solid red line, which is attached to the belly of the orange ball where it represents the trajectory of the signal. Yellow arrow tips specify the respective direction of movement. Approximately twelve dark purple balls appear all across the rendered field in order to represent obstacles that can affect the trajectory of the signal. Rose cylinders, finally, represent the cytoskeleton, which is expected to put limits on (and perhaps directing impacts to) the protein's trajectory. As regards the *differences* between the screenshots, the inscription indicates that the shots "[b]" – "[e]" vary in graphical terms. Accordingly, "[b]" accords darker colours to those parts of the cytoskeleton, that are to appear further away than others, "[c]" adds "depth of field" (p. 171), so that items that are further away loose acuity; "[d]" adds "depth cues"

(p. 171), that is, desaturated colours to those parts that are not in the focus. Finally, "[e]" combines all measures, so that a fore- and background is defined and entities of interest are highlighted. In practice, researchers can choose between these measures and thus define the focus.

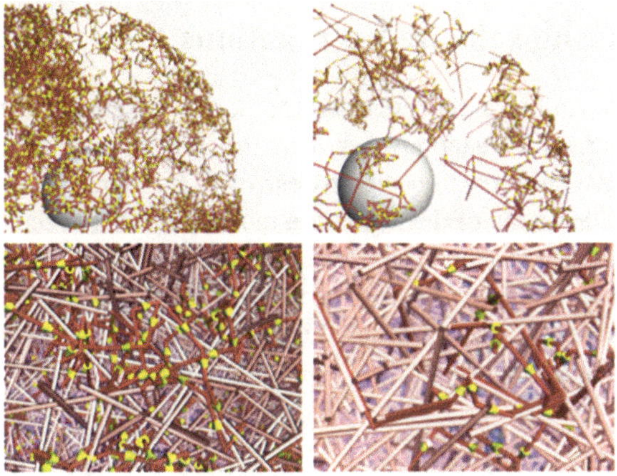

Fig. 22: Modes of signal transport (cf. Falk et al. 2009, p. 172)

Close-up analysis of Fig. 22. Like Fig. 21, Fig. 22 presents a *set* of screenshots, which is used to visualise the trajectory of a protein. This time, the screenshots are arranged differently; instead of one horizontal row, they form two rows of two items, respectively. What is more, this time there is no geometrical, topological or graphical congruence between either of the compiled items. Instead, the latter share some perspective-related aspects, such as a bird's-eye view on the upper-right quarter of the modelled cell (top row) and a zoom into the cell interior (second row). Moreover, the two screenshots on the left-hand side and the two screenshots on the right-hand side belong together in the way that they present one and the same scenario from different zoom levels. Each of the sceneries exemplifies a different type of protein movement. The scenario on the left side is set in correspondence with the type of particle movement that works via *diffusion*, whereas the scenery on the right-hand side is dedicated to the type of movement via "*motorized transport along the cytoskeleton*" (p. 172). In accordance with the biological reference model, the movement via diffusion appears to be less directed than the movement via transport. In both cases, however, the signalling molecules seem to cross the walls of the cytoskeleton, which is due to the fact that "only every thousandth position is visualized" (p. 172), meaning that this – actually unintended – effect is accepted for the sake of a feasible computation. Accordingly, some rendering measures foster the tractability rather than the adequacy of the model. For the same reason, "the lower half of the cell was cut away and the edges of the cytoskeleton are emphasized for better distinction in the lower left image" (p. 172). In other respects, Fig. 22 is build along the lines of Fig. 21, which is why the key to the glyphs is not repeated, meaning that, once established, it holds for every extraction of the source figure.

3.2.6 Rendering scheme for reactions

The authors also aim to represent mechanically describable interactions and chemically describable reactions between the cell components, because these events could have an impact on the process of signal transduction. For example, it might happen that none of the signalling molecules reaches the cell core (which is however necessary to accomplish the signal transduction), because they get deflected due to a collision with other particles or with the cell tissue, or because they disappear due to chemical decay. Fig. 23 exhibits the cues, which have been chosen to indicate such events.

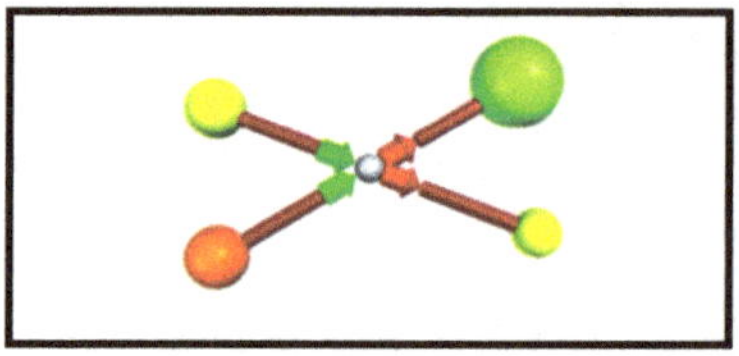

Fig. 23: Coding scheme for reactions (cf. Falk et al. 2009, p. 171)

Close-up analysis of Fig. 23. Fig. 23 specifies the glyphs that have been chosen to represent interactions between molecules. The glyphs are not presented in the context of the whole cell, but they are isolated and simply illustrate themselves, that is rather to say, they introduce the corresponding coding scheme. The rules of combination and interpretation are briefly defined as follows: The "[r]eactions (grey) between molecules are illustrated by arrows. Reactants (green) enter the reaction and products (red) leave it" (p. 171). Accordingly, the inscription focuses on the arrow glyphs only. The balls and lines (though coloured and therefore graphically equipped) are just variables for possible reactants, products and routes.

3.2.7 Tractability measures: cuts and transections

As mentioned above (cf. 3.2.5), additional rendering facilities are implemented in order to promote the orientation in and navigation through the cell model. This is needed since "when looking at events deep inside the cell, crowding objects often hide the objects or events of interest. Although transparency allows [us] to look through the obstacles, it creates a haze of shapes that [...] disorients the eye" (p. 171). Fig. 22 (analysed above) is again referred to in this context, as it helps demonstrate that completeness (in biological terms) is not the number one priority in any case. Instead, "it is much more helpful to cut away the perturbing parts of the cell to allow a clear view at the object of interest" (p. 171). Another advantage of cutting parts away is that the pieces can be conjoined in such a way that a "transection" (p. 171) through one or more cells or cell layers is mimicked. Accordingly, the biological reference base does not only include models of the cell interior, but also empirical research methods, such as "wet lab experiments" (p. 169). The dedicated renditions are meant to be used when it comes to simulating the diffusion of drug molecules through solid tumour cells (cf. Fig. 26; cf. 3.2.11).

3.2.8 Tractability measures: depth cues

As has been emphasised on several occasions, the reduction of the model's complexity presents a key challenge to the authors. Due to the high number of ingredients and their differences in size, the cell interior has the air of a "crowded environment" (p. 171). Nevertheless, all elements are involved on purpose. The cytoskeleton, for instance, helps derive topological information about the otherwise context-deprived "point cloud of proteins" (p. 171). In order to foster this effect, further "focus and context techniques" (p. 171) are employed (cf. close-up analysis of Fig. 21). These include variations of the colour gradient. The nucleus, for instance, is given a colour which is perceived as *cooler* than that of the surrounding elements. Thanks to this modification of the colour code, "the user is always aware in which direction the camera is aimed" (p. 171). The techniques also include "*depth of field*" (p. 171), that is, the definition of fore- and background by means of a reduced acuity of background parts. Finally, *colours are accentuated* depending on the location of the coloured items. In particular, "the more an object is out of focus the less saturated it is" (p. 171). All measures listed are exemplified, using Fig. 21 (see above).

3.2.9 Alternative rendering: microscopy-like images

In laboratory research on intracellular signal transduction, microscopic images are extensively used. Therefore, the authors attempt to render the computed data in a "similar manner" (p. 171), in addition to the geometrical rendition. Corresponding reference models are taken from "confocal laser scanning microscopy" (p. 171), where a laser beam is used to excite fluorescent dyes that are tagged onto proteins. "The focus of the laser beam is very small (250nm×250nm×750nm)" (p. 171) and is only successively aimed at single parts (in this case tiny points) of the primed cell tissue. An apparatus detects the emitted light and, thanks to a small aperture, helps increase the resolution of the entering spots. The latter are assembled into a picture by the aid of a computer (cf. p. 171). In order to indicate how the results look like, the authors present an example at the beginning of the paper (cf. Fig. 19 [left]). The task now is to render the simulation (resp. its solutions) in a way that comes close to this. To this end, the authors "apply a radial intensity distribution to each protein to mimic the Gaussian intensity profile of the laser" (p. 171f.). The effect of this is that the proteins appear brighter in the centre and dimmer at the corners, so that, thanks to an additional "blending function" (p. 172), the contour of the proteins seems to fuse with its surroundings. In accordance with the reference images, the surroundings are not further defined; only "[b]ackground fluorescence is imitated" (p. 172). The result is "an image with high similarity to the experimental ones, thus allowing an easy comparison" (p. 172). This comparison is made possible for the reader of the paper as well, thanks to the introductory sampling of images (cf. Fig. 19).

3.2.10 Application I: transduction types for MAPK

After having introduced the biological architecture and two rendering schemes of the cell model, the authors report two "application[s]" (p. 172). That is to say, they tackle two biological problems with the aid of the model.[50] The first problem concerns the "question whether signalling molecules (esp. "mitogen-activated protein kinase") diffuse through the cell or [whether they] are transported [by] motor proteins along the cytoskeleton" (p. 172). Fig. 24 illustrates the corresponding biological premises that are taken into account in order to find an answer to this first question.

The authors stress that the results of the simulation can be compared with the results of experiments on cellular signal transduction. Moreover, they maintain that the empirical accounts are supplied with new hypotheses on the basis of the results of the simulation. In particular, according to the simulation, "diffusion

50 The authors of the other paper, examined earlier, made use of a similar strategy (cf. 3.1). However, personal communication revealed that this approach is considered as an *application* in computer science, that is, as a technical test case and concurrent service to biology. In biology, on the other hand, the approach is considered as a contribution to *basic research*, that is, to the very core of the research program.

leads to [a] radial gradient of active signalling molecules in the cell, [whereas] the transport with motor proteins [...] increases the concentration around the nucleus" (p. 172). Fig. 22 and Fig. 25 support this reading of the results. Accordingly, in the case of diffusion, proteins tend to move farther per time step, whereas in the case of motorised transport, the distribution of proteins is less disturbed on their way to the nucleus. These findings are used to derive the hypothesis that "diffusing particles can reach the nucleus faster than those transported along the cytoskeleton, but the probability of doing so decreases strongly with the distance" (p. 172).

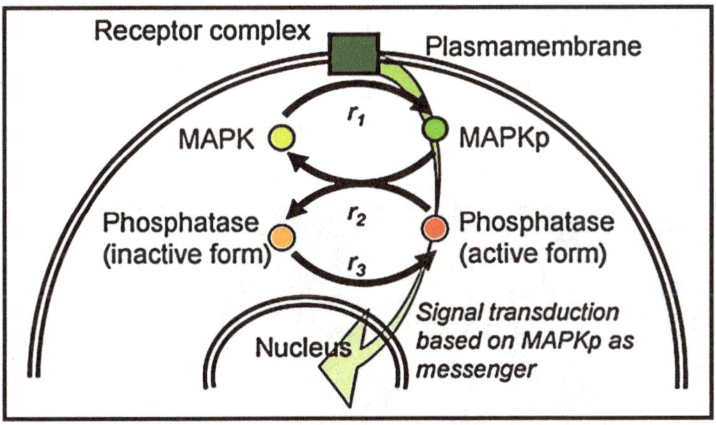

Fig. 24: Signal transduction mechanism (cf. Falk et al. 2009, p. 172)

Close-up analysis of Fig. 24. Fig. 24 builds upon a set of geometric primitives (i.e. lines, arrows, quadrates and points), a colour code and a set of inscriptions for each geometric ingredient. The largest of all elements is a black double line that forms a semicircle and that is used to indicate the cell's membrane. The cell interior is filled with different items, all of which contribute to the illustration of the signalling mechanism. A dark green square is located on the membrane, so that half of it reaches the interior and the other half remains outside the cell. In biological terms, it is meant to represent the "receptor complex" (p. 172), that is, the very point or "scaffold" (p. 172) where external signals are received and transmitted to the cell interior. A light green arrow, reaching from the receptor complex to the nucleus, indicates the pathway of the transmitted signals and thereby the direction of reading; accordingly, the arrow is used to point out that the distinct stages of the transduction process are listed chronologically from the top (receptor complex) to the bottom (nucleus). The process begins with the arrival of a signal at the reactor complex where it stimulates an initial reaction ("r1"), in the course of which the mitogen-activated protein kinase (abbr. "MAPK") is changed into the activated, phosphorylated "MAPKp" (p. 172). The latter is assumed to be the "mobile" (p. 172) component of the signal transduction; the arrival of

MAPKp at the nucleus thus completes the whole process. However, it is not guaranteed that it truly reaches the nucleus, since in contact with the enzyme "active phosphatase" (p. 172), MAPKp retransforms into its non-phosphorylated form ("r2"). In this case, the enzymes also relax back into their non-activated form ("r2"). The activation of these enzymes ("r3") may thus help influence the signalling strength, given that the latter depends on the number of MAPKps reaching the nucleus. This much is said about the biological reference account. From the visualisation point of view, the mechanism is particularly interesting, because "[t] he local excitation (at the plasma membrane) together with the global inhibition [...] leads to spatial gradients" (p. 172). In other words, due to its spatial components (as involved, e.g., in the description of the particle movement), the process (resp. its biological and mathematical account) is suitable to be rendered by virtue of geometrical, topological or graphical rendering schemes (as proven with respect to Fig. 21 and Fig. 22). One could thus argue that Fig. 24 does not only illustrate the model of the biological mechanism under consideration, but it also helps justify the efforts invested in the visualisation of the simulation-based approach. Fig. 24 thus serves at least two functions in the paper.

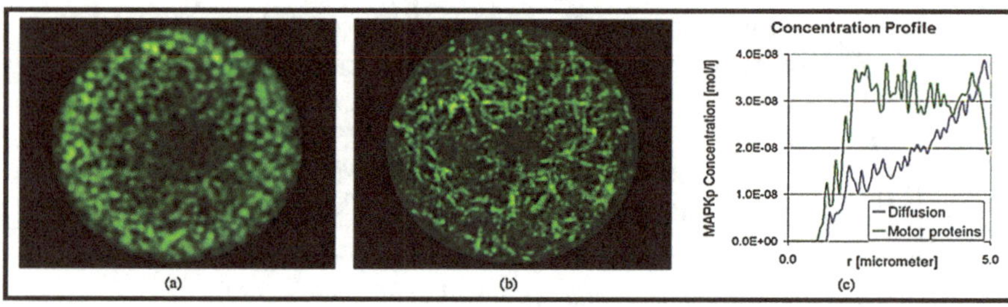

Fig. 25: Differences in protein concentration and crowding (cf. Falk et al. 2009, p. 173)

Close-up analysis of Fig. 25. This set of figures is mentioned several times in the paper. It is first mentioned when it comes to defending the glyph-based rendition from a more general, methodological point of view (here: cf. 3.2.2). It is argued in this context that the graphs usually employed (cf. Fig. 25 [c]) can only allow for an average number of molecules per radial gradient, and that the graphs also do not provide any information about the individual positions of the molecules and their distances to each other, which implies that graphs cannot encode phenomena such as particle crowding. Having thus defended the idea of alternative rendering schemes, Fig. 25 is again referred to when it comes to introducing "microscope-like images" (p. 169) (cf. 3.2.9). In this case, however, the reference is only *indirect*, since it is actually Fig. 19 that is mentioned, and Fig. 19 only alludes to Fig. 25. Another, more *direct* quote can be found in the section that presents the application cases (cf. 3.2.10). Here, the figures (esp. Fig. 25 [a] and [b]) help corroborate the claim that there is an optically discernible difference between the signal transduction via diffusion (left) and the signal transduction via active motor transport (centre). It is then argued that this difference could help "clarify whether the respective signalling molecules are actively transported through the cell in vivo" (p. 172). In order to foster the comparison of the simulation results with those of experiments, both figures are presented in the "microscope-like" (p. 169) rendition. This rendition is dominated by a glaring green colour, which is meant to imitate the fluorescent light emitted in the course of confocal laser scanning microscopy. The green light accumulates in a circle that indicates the cell contour, and it reveals gradual differences in brightness. According to the differences in transduction revealed before (cf. 3.2.9, 3.2.10), the brighter green parts form blurry dots in the case of diffusion [a] and less blurry scuffs in the case of active motor transport [a]. On this basis, this set of figures may serve as a representation of the computed results and possibly also as a guide to the interpretation of real "microscopic images" (p. 171).

3.2.11 Application II: delivery of drug molecules

The second domain that is expected to profit from the 'application' of the simulation (resp. its rendition) is research on the "delivery of drug molecules to cancer cells in tumours" (p. 172). This process exposes some traits of the previously described transduction mechanisms, insofar as drug molecules also work as signals when they "trigger the death of the cancer cell" (p. 172). Moreover, drug molecules also have to pave their way through a "crowded environment" (p. 169). On the other hand, a "tumour is much larger compared to single cells" (p. 172). This presents a challenge to visualisation, since "[t]he smallest object is not visible when looking at the tumour while the spatial dimensions of the tumour cannot be sensed when focusing on single drug molecules" (p. 172; cf. Fig. 27). The authors thus go for a small tumour, which also has the advantage of reduced "computational complexity" (p. 172). Despite the remaining problem of balancing scales, the authors insist that, thanks to its elaborate rendition, the simulation can be adapted to the second problem, too. That is to say that empirical accounts are supplied with the option of comparing simulation and experiments in search of further hypotheses.

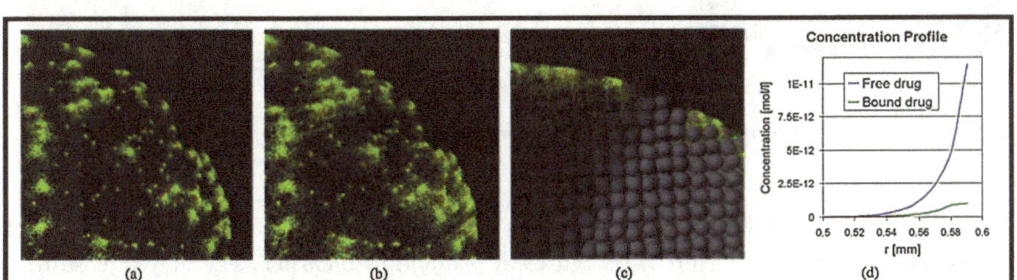

Fig. 26: Drug delivery to cancer cells in a tumour (cf. Falk et al. 2009, p. 173)

Close-up analysis of Fig. 26. This set of figures is part of the second application of the "virtual cell" (p. 169) and the related methods of computation and rendition. The idea is to apply them to the delivery of drug molecules to cancer cells in a tumour. From the viewpoint of visualisation, this task is challenging, because a tumour consists of many cells, and because the signalling particles are very small compared to the entities that form its (functional) environment. The visualisation is thus faced with a multi-scale problem, that is, the process is best described on *different* scales. Fig. 26 [a] – [c] exemplifies the suggested compromise of the authors, which can be consulted in addition to the conventional graphs presented in Fig. 26 [d]. By way of compromise, the authors accordingly present only a tiny part of the tumour (seemingly the upper-right-hand corner), which helps discriminate both the single cells, presented as grey balls (cf. Fig. 26 [c]), and the moving drug molecules, marked by tiny yellow-greenish dots (cf. Fig. 26 [a] – [c]). The colour of the signalling particles is thus varied, compared to Fig. 25, probably in order to avoid the confusion of the two application cases. Fig. 26 [c] (centre right) furthermore helps discern the differences in proportion, since, by outlining their shapes, it allows for estimating the high number of cells that is focused on (here: approx. 80). Apart from these measures in support of optical tractability (recall that Fig. 25 [a] and [b] focuses on *one* cell only), the representation is intended to foster comparisons. To this end, the cancer cells (presented as grey balls, partially shaded and ordered in parallel) are rendered in a way that mimics the looks of a 'transection' as conducted in laboratories (cf. 3.2.7). On these grounds, a biological hypothesis is derived, namely that drug delivery proceeds relatively slowly from an observer's viewpoint; 30 minutes after injection, the drug particles still crumble at the tumour's surface. This is particularly argued for with regard to Fig. 25 [a] – [c].

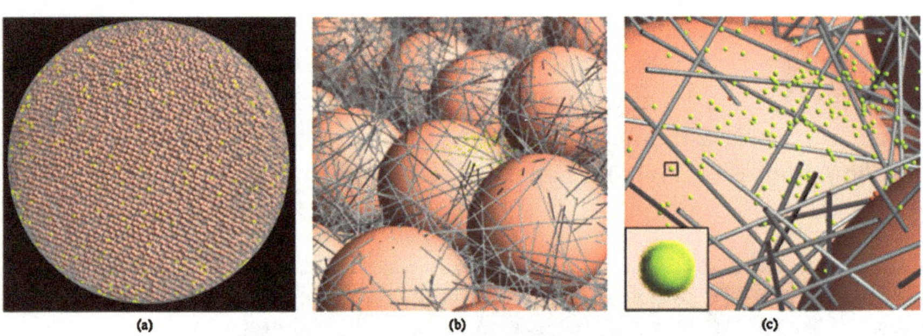

Fig. 27: Size of tumour and drug molecules (cf. Falk et al. 2009, p. 175)

Close-up analysis of Fig. 27. This figure is actually the last one presented in the paper. However, as with some of the other figures, it is mentioned earlier. One may thus surmise that it simply could not have occurred earlier, given that the authors prefer to locate such triples in the upper half of a page, but never *before* they are mentioned in the text, which puts limits on their possible locations. In fact, Fig. 27 also belongs to the second case of application (cf. 3.2.11). In this context, it is used to give an idea of the conjoined levels or scales of description. As has been stressed before, the latter are difficult to combine, so that it remains impossible to present all model components in one shot. The latter must indeed be introduced successively. Fig. 27 [a], for instance, provides a distance shot of the whole spherical tumour (1,2 mm). The distance shot builds upon the rendering scheme of the previous application insofar as it exposes roughly the same shape of what has been introduced as a single cell before (cf. Fig. 25). Moreover, it is (likewise) adapted to laboratory upshots, since tumours in vivo are not perfectly spherical items. And the round black frame (Fig. 27 [a]) indeed reminds one of a microscope that is focused on the item inspected. However, the most salient aspect is that it becomes obvious that the total view does not allow for discriminating smaller units, such as drug particles, which is why it makes sense to implement a zooming option. From a phenomenological point of view, this comes at the cost of comprehensiveness, since not all items can be rendered sharp at each zooming level. Yet, since the model is meant to supply researchers with spatial information about the signalling process in the first place, this deficit can still be compensated by the suggested measures.

3.2.12 Technical aspects

After having introduced two applications, the authors present additional information on the modelling assumptions as well as on the techniques combined.

As regards the *simulation*, the authors stress again that they have assumed a difference between mobile and immobile parts of the cell. Only the mobile parts are programmed to behave either in accordance with diffusion, which is conceptualised as a "random walk" (p. 173), or with motorised transport, which is conceptualised as a movement along the "cytoskeleton fibre" (p. 173). Further idealisations are aligned with data and experiences drawn from experiments. For instance, it is determined that, "[i]f agents collide with fixed objects, the step is not carried out" (p. 173). Furthermore, reactions can only occur to an extent that "fit[s] the macroscopic reaction rate" (p. 173).

Thanks to the visualisation techniques offered in support of the analysis of the simulation (resp. its solutions), the authors enable researchers to modify the rendition interactively. Accordingly, users can choose, modify or manipulate their own focus and the focal areas, and they can highlight aspects of potential interest (e.g. by changes of the colour code and shading; cf. e.g. 3.2.8). To this end, the computation of the graphics is delegated to the graphic processor and thus separated from the solution of the stochastic model, which is computed by the central processing unit. The drawback is that the user needs to wait for the results of the simulation. The advantage, however, is the fostering of interactivity (see also below; cf. 4.3.3.3). On top of that, the user is enabled to experience her/his changes as "a smooth transition from the old value to the new" (p. 174); thus s/he receives an optical feedback and may evaluate the wins and losses of information on that score.

The authors also add some further information about the implemented mode(s) of *interaction*, emphasising that the rendered cell model can be inspected successively using keystroke commands. This facilitates the inspection of the model along the lines of a self-guided journey through the "virtual cellular environment" (p. 175). In particular, the user is, for instance, able to skip "frames" (p. 174), to hide information, such as occluding cell material, and to prepare "slices" (p. 174) as if they were prepared for or in the course of microscopy. Furthermore, the user is free to choose the zoom factor as well as the coding "style" (p. 174). Thus, the idea is that visualisation is an explorative process rather than a manifestation of pre-prepared interpretations of the computed data.

3.2.13 Results

The section entitled "results" (p. 174) provides detailed information on the conditions that led to successful applications (cf. 3.2.10, 3.2.11). "Table 1 (here: Fig. 28) shows the number of components computed in the simulations for the MAPK and Drug Diffusion data sets" (p. 174). Fig. 28 thus corroborates two aspects, namely the size of the model and the similarity between the two cases of application, which build upon analogous, systematically relevant traits (e.g. insofar as they employ three types of components of which some can move). By contrast, Table 2 (here: Fig. 29) adds information about the computational performance of the proposed rendition. The performance is measured in "frames per second" (p. 174), meaning that one counts the number of consecutive images that are computed per second. The performance is measured separately for different perspectives ("Close-up" and "Complete"), rendering units or "instances" (such as proteins and obstacles; cf. 3.2.5), "lines" (such as cytoskeleton fibres; cf. 3.2.5), rendering schemes ("depth of field") and applications ("MAPK" and "Drug Diffusion") (p. 174). In the case of drug diffusion, certain memory units on the graphic processor (i.e. "vertex buffers" ["VB"]) are switched off (cf. p. 174). This implies that there is *no* direct interaction possible, which is ultimately to the benefit of this highly resource-consuming application (cf. Fig. 29). Still, the process takes time in both cases; in fact, "[s]even seconds of the MAPK simulation took about 12 hours" (p. 174), varying with the number of involved objects and the "chosen time step" (p. 174). By contrast, "the data set shown in Figure 7 (here esp. Fig. 26 [c]) took nearly a week for computation" (p. 174).

MAPK	
Proteins	70,000
Obstacles	43,062
Cytoskeleton Filaments	6,097
Drug Diffusion	
Drug Molecules	143,634
Cells	107,303
Inter-cellular Fibers	887,818

Fig. 28: Number of components per type (cf. Falk et al. 2009, p. 174)

Close-up analysis of Fig. 28. This table builds on a minimal set of two horizontal rows for each of the two application cases. The first row is identified by the name "MAPK" (p. 174); the second row is devoted to the second application and has the title "drug diffusion" (p. 174). The authors have refrained from adding any vertical lines and guide the eyes only by means of the parallel listing of the entries.

MAPK					
	Dof	VB	Instance	Lines	None
Complete	yes	32.00	31.94	45.00	59.22
	no	51.77	51.70	101.57	145.36
Closeup	yes	34.37	34.25	46.24	58.94
	no	58.85	58.64	124.21	145.64
Drug Diffusion					
		VB	Instance	Lines	None
Complete		–	1.87	8.03	16.22
Closeup		–	2.98	9.17	9.93

Fig. 29: Rendering performance (cf. Falk et al. 2009, p. 174)

Close-up analysis of Fig. 29. Table 2 (here: Fig. 29) provides information on the rendering performance, measured in "frames per second" (p. 174). The results are presented separately for each application case. Most of the categories listed in the first line of each index have been clarified above (cf. 3.2.13); it remains to add that the abbreviation "Dof" (p. 171) standing for "depth of field" (p. 171), which is, judging by the entries, highly resource-consuming. In this connection, it becomes apparent that the tables allow for a quick grasp of the values, since each category is only mentioned once; yet the types and relations of the entries can still be derived from the way they are sorted and related to each other. This effect is fostered by the proximity of the two tables, which makes it possible to compare their entries and, on this basis, support comparative claims about the efficiency of computation.

3.2.14 Conclusion and future work

The authors end with a brief summary and outlook. Their major objective was to propose two auxiliary schemes and devices for rendering the solutions of a stochastic, multi-agent simulation of signal transduction. One of them, the "*schematic illustration*" (p. 170), scores well as an aid to heuristic approaches, also since it allows for an "interactive segmentation and selection [of areas]" (p. 175). The other scheme, the "*microscope-like*" rendition, is in turn "easier to grasp than the detailed schematic representation" (p. 175), and it can be better compared to confocal laser scanning microscopy. Making use of these capacities, both schemes are applied to two cases from "ongoing research in the fields of biology and medicine" (p. 175). Both applications are deemed successful – not only, but also thanks to the detailed setting, which can "justify the high computational efforts" (p. 175). The authors want future accounts to focus on the coupling of simulation and visualisation and the improvement of computational tractability and biological adequacy.

3.3 Abstractions

Having introduced two series of figures (as well as three tables) by way of describing, re-narrating and re-contextualising their uses as expounded in the consulted journal papers and personal communication with the authors, the remaining task is to tie up the threads and to set demands on a conceptual account of picture-like means in science.

3.3.1 Functional involvement of figures

First of all, it turned out that there are various ways to use figures and that one and the same figure can be used in various ways. These two findings make it seem difficult to capture the nature and role of figures in terms of *mono*-functional accounts (which are still found debated; cf. 4.1.3). One may rather speak of a nesting of means-ends relations, considering that figures or parts thereof serve different functions simultaneously.

These functions concern **(1)** the *compositional* level (e.g. "trajectories connect the sampled points with straight lines", Falk et al. 2009, p. 170), **(2)** the *representational* level (e.g. "As seen in Fig. 5(c) [here: Fig. 15], first applying z_O slightly increases x_O", Schittler et al. 2010, p. 6) and **(3)** the *methodological* level (e.g. "The outcomes of scenarios (a) – (d) are depicted in Fig. 5 [here: Fig. 15]", Schittler et al. 2010, p. 5).

Each level imposes criteria of adequacy on the design and treatment of the figures. For example, their lucidity is required on the compositional level (i.e. the possibility of discerning the graphical components of a figure), semantic coherency on the representational level (i.e. the possibility of pairing the compositional components with the [e.g. mathematical or biological] reference concepts, while preserving the relations between these concepts as defined by the prior renditions [e.g. descriptions] of the reference system). Last not least, there is significance expected on the methodological level (i.e. that a figure is useful to the accomplishment of a [e.g. scientifically or aesthetically justified] task).

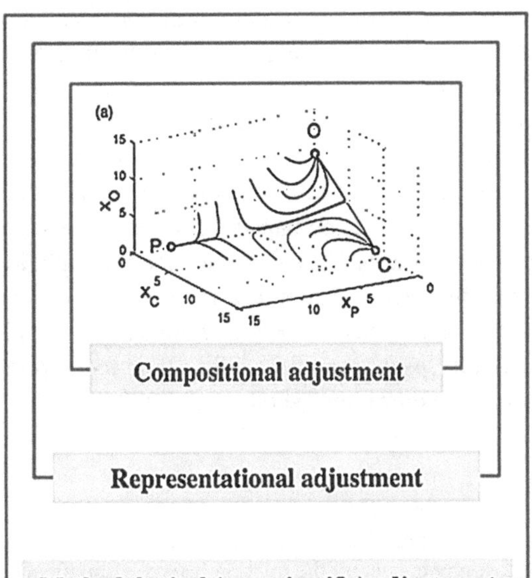

Fig. 30: Levels of adjustment

Some levels require adequacy on other levels, too. The representational capacity of a certain figure may, for example, depend on how detailed the composition can be made without losing its optical or computational tractability (cf. Falk et al. 2009, p. 175). The procedural significance may, in turn, depend on the representational capacity, for instance, in the sense that a comparison of simulations and experiments requires a successful matching of what is represented by or via each of them (cf. Falk et al. 2009, p. 169). Then again, in order to give sense to phrases such as "non-representational painting" (cf. e.g. Elpidorou 2010; Giere 1996, p. 272), one may assume that instances of that category are approached with regard to compositional aspects only, or taken as representations of merely compositional schemes and patterns, like colour combinations or brushstrokes.

Hence, by any means, there is discursive support for the adding as well as stripping away of functional embeddings, which is why it seems appropriate to maintain that there are, indeed, distinct frameworks that offer combinable requirements or 'programs' (cf. 2.2.1.2) for the design and use of figures. This being the case, recent accounts would not only be expected to capture the different levels of functional involvements (cf. Fig. 30), but one would also want them to account for the relations and dependencies between these levels without suggesting resolute determinacy. Determinacy is not an option worth considering, if you think of all the inscriptions and allusions that are attached to the figures and that would be entirely dispensable, if there was some kind of invariant interconnection between the levels of composition, interpretation and further use.[51]

In this context, the analysis also revealed that each figure is linked to *another* source of information (e.g. to a file of numbers) which is made present by the figure in a certain way or mode, perhaps with a certain loss, addition or distortion of the conveyed information. The so-established *relata* allow for further uses germane to the scientific program they are involved in. For example, they help to "derive and support assertions" (close-up analysis of Fig. 12), to "sum up the model components" (close-up analysis of Fig. 9) or to "support the comparison between simulation and experiment" (Falk et al. 2009, p. 171).

51 The reason why there must be a comment on determinacy at all is that there were attempts in philosophy to specify the relation between graphical compositions and the things 'seen in' or 'seen via' them in terms of structural correspondence (cf. 2.3.1). However, approaches like that apparently conflict with the mere resources that are spent on the clarification and incorporation of the figures.

Following this repetitive scheme amongst the reconstructed procedures, I suggest differentiating between 'figure-like' and 'figure-based' means, and also – if applicable – between 'picture-like' and 'picture-based' means.

With this, the term 'picture-like' could still be attributed to instances of geometrical, topological or graphical coding schemes (cf. 2.2.2.2), but with the addition – derived from their implementation as scientific figures – that these instances contribute to the linking of different sources of information. The term 'picture-based', which I propose in addition, would then denote further involvements that pose additional context-variant and error-prone[52] demands on the establishment and treatment of these means and measures.

3.3.2 Practical implications of figure-likeness

However, it is not only the multitude and interconnectedness of functional involvements or 'programs' (cf. 2.2.1.2) that should be allowed for by current accounts of figure-like and perhaps also picture-like means in science. The foregoing analysis of examples also corroborates another intuition, which may help link the studied practices (resp. the studied testimonies thereof) to the recent efforts of conceptualisation, and vice versa.

In particular, it seems to be the case that the traits, which qualify something as figure-like, and possibly also as picture-like, result essentially from certain *acts*, that is, from the establishment and treatment of composite entities *as* figures of certain kinds and capacities. One might even go as far as to say that these acts *are* the relevant instances of these means, which need to be consulted when it comes to tackling conceptual controversies (cf. Seja 2009). This latter thesis is actually well supported.

First of all, one can maintain that the notion of *means* is not restricted to a certain type of entity. Operations, methods, institutions, etc., can all take the part of means in means-ends relations, given that this is the defining condition (cf. Hughes, Kroes & Zwart 2007). On this assumption, it is at least *not impossible* that certain acts *are* instances of figure- and picture-like *means*.

---

52 It is a matter of controversy whether the instantiation of 'picture-like' entities can also fail and result in something like a non- or not-yet-picture-like tool (cf. Richtmeier 2009, p. 140ff.; cf. Wittgenstein *Note-books 1914–1916*, 26.11.1914). The sheer puzzlement about this issue and, even more, the rejection of the notion of *failure* for the first category (at least by both of the named authors, which is not easy to decide in the case of Wittgenstein), could be seen as a further support for the above-proposed distinction between picture-like and picture-based procedures.

Moreover, one can hardly deny the impression that the implementation of visual aids is a not very easily framed *process* (cf. Boehm 2001, p. 53). As a consequence, it does not make much sense to dedicate the attributes 'figure-like' or 'picture-like' to *what comes out* of the respective pipeline of operations, for the plain reason that it is not evident, nor does anyone claim, that the associated series of acts (e.g. highlighting, zooming, etc.) *must have* a definite end and outcome (cf. Falk et al. 2009). On the other hand, it also seems wrong to avoid the notion of a process by arguing that *every single* screenshot or eye blink captures a veritable instance of figure- or picture-like means. This simply collides with the fact that rules of combination and interpretation are introduced *once* for whole sets and even variations of optical inputs (cf. close-up analysis of Fig. 21). A steady trait thus rather seems to be the *enforcement of certain coding schemes* (e.g. the use and combination of different [e.g. graphical, geometrical or topological] descriptors; cf. 3.1f.). This would also fit the notion that one and the same set of correspondingly describable marks is treated as 'picture-like' on one occasion and as 'non-picture-like' on the other, depending on various factors, such as the familiarity with certain coding schemes (e.g. with hieroglyphs) or the wish to test-run a variety of those schemes. Hence, it seems that even the consequences of a practical notion of 'figure-' and 'picture-likeness' catch up with empirically inspired intuitions on the matter.

Last not least, a practical account would also comply with the state of research, since one tends to speak of "operative scriptures" ("*operative Schriften*", Fischer 1997, p. 93) or "operative symbolism" ("*operativer Symbolismus*", Krämer 1988, p. 176) in the context of computer-aided coding systems, stressing their openness to and dependency on context-adapted configurations on different levels (cf. Fig. 30). Similar views also shape the more general debate about the nature of pictorial means (for an exemplary overview; cf. Seja 2009). Here, one also tends to single out specific *operations* as the defining or co-constituting instances (e.g. acts of "framing" or "fencing" ["*Abzäunung*", Richtmeyer 2009, p. 143] areas of intended picture-likeness or the procedure of semantic disambiguation via perspective-bound "gaze[s]", Alloa 2010, p. 14). Thus, the focus on practical implications of figure- or picture-likeness seems promising from recent perspectives (until further notice; cf. 4.1.1).

3.3.3 Methodological relevance of figures

It is clear from the foregoing section that it makes good sense to allow for the idea of practically established and entangled frameworks that guide the implementation of picture-like means in science. A closer look, however, reveals that this idea also illuminates the relation between different functional ascriptions *on one and the same* level.

Consider, to start with, an independent example, the "disappointed chimpanzee" by Ch. Darwin (1872, p. 141; Fig. 31). One may well introduce this sketch as a *depiction.* In so doing, one qualifies the sketch as a *guide to characteristic features* of disappointed chimpanzees, perhaps hoping to recognize such specimens on future occasions by virtue of the sketch (the corresponding notion of *depiction* is coined by Kötter 2007, p. 162ff.). On the other hand, one could use the very same sketch as a *documentation* – that is, as a *testimony that helps to gain a deeper conviction about the truth of certain (claimed) facts* (cf. Kötter 2007, p. 168); for example, the claim that sketches aided the depiction of animal behaviour in the 19th century. Similar entanglements of mutually reliant usages are described in Lynch (1988, p. 203).

Accordingly, it seems appropriate to buy into the premise that it is possible (technically speaking) to *change* or *merge* different methodological involvements of figures.

In fact, the consulted journal papers (Schittler et al. 2010; Falk et al. 2009) only confirm this hypothesis, for they reveal similar entanglements in the functional embedding of the employed figures.

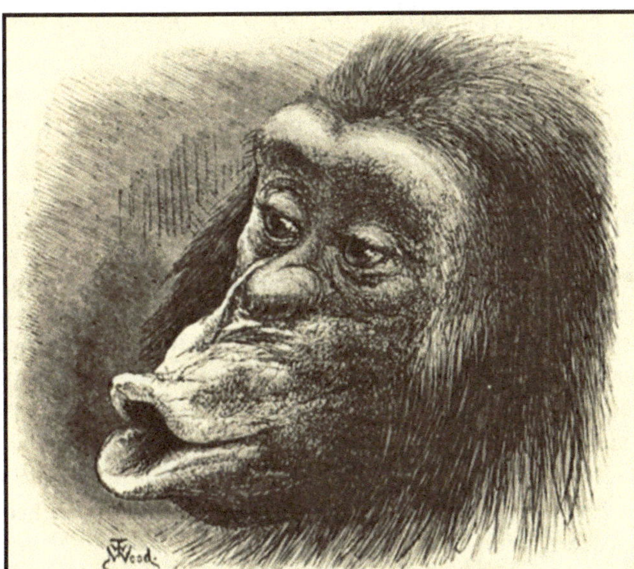

Fig. 31: Disappointed chimpanzee (cf. Darwin 1872, p. 141)

For instance, Fig. 25 [b] occurs already in Fig. 19 (centre left). However, while Fig. 25 [b] is meant to display the characteristic features of the "microscope-like" (Falk et al. 2009, p. 169) rendition of simulation results, the same screenshot is used as an *illustration* (i.e. as an *aid to the consumption of descriptions or [other] accounts*; cf. Kötter 2007; p. 176) of the combined approaches and fields of expertise in Fig. 19. Another, perhaps more subtle switch of the functional embedding concerns, for instance, Fig. 13, which is used (before publication) as a *visualisation* of model solutions (i.e. as a *more tractable rendition of numerical data*; cf. Kötter 2007, p. 171ff.). Then, after slight changes in the composition (cf. Fig. 14), the figure is additionally used as a *rationale* for the proposed interpretation of the model solutions (cf. close-up analysis of Fig. 14).

In this context (esp. also in regard to the taxonomy of Kötter 2007), it is worth adding that the methodologically justified embeddings of the figures are not confined to the repetitive or confirmative schemes of use, which draw mainly on the figure's capacity to relate to other bases of information (e.g. convolutes of sentences or numbers, or certain claims; cf. 3.3.1). There are also cases in which figures reveal *unique* capacities – not in the mode of repetition or transcription (cf. 3.3.1), but in the mode of *suggestion*: Fig. 15, for instance, allows for the imaginative adding of vertical lines, which facilitate the comparison of the vertically aligned graphs. In fact, the whole second approach by Falk et al. (2009) is devoted to the exploitation of exactly this capacity, namely the *intervention into the rendition* (which is left to one's imagination in the first case; cf. close-up analysis of Fig. 15). Falk's (2009) approach is particularly devoted to securing and extending this intervention as a fixed option, in order to allow for further usages – especially for the 'interactive' exploration of a model's solutions (cf. Falk et al. 2009, p. 174; cf. 4.3.3.3).

3.3.4 Compositional and representational aspects of figures

The previous paragraph indidcated that analytical accounts are faced with the problem of defending different types of usages against the background of entangled practices. Yet this is more of a challenge than a dead end, because analytical approaches are not expected to *imitate* the entanglement that practice is marked by as an integrated whole. One would rather want them to propose distinctions (cf. 3.3.3) – without suggesting that the concepts thus distinguished correspond to separate or closed systems in practice.

In this paragraph, I would like to discuss another problem that might also be a challenge to conceptual endeavours, namely the question: what kind of entities are actually used in the afore-noted ways (cf. 3.3.3)? In other words, if it is true that 'picture-likeness' somehow emerges from using or approaching something in a certain way (cf. 3.3.2), the question naturally arises whether there is anything to specify about that *something*.

In regard to this question, there could not be more discrepancy between theory and practice. While theory (esp. philosophical theory) has for long been preoccupied with the *conveyors* of coding systems more or less in general (e.g. under the common phrase of 'media'; cf. e.g. Mersch 2006a, Hoffmann 2002), scientists postulate the common ability to discern and exploit different devices of signification or representation *as such* (even more as their use is an initial part of the inauguration in the targeted scientific communities; cf. Kötter 2007, p. 161). In practice, the main concern is rather to guarantee the fast and faultless assimilation of a code and the information thereby encoded.

The most prominent measure in this context is the use of *inscriptions*, which provide keys to the combination of compositional units (e.g. circles; cf. Fig. 12) and conceptual correlatives (e.g. equilibrium points; cf. Fig. 12). It seems, though, that the inscriptions (and other parts of the context) cannot facilitate the functional analysis of each figure and its components to the same degree. Fig. 9, for instance, can be *fully* divided into discrete parts, given that the colour white is (in this context) taken as an unoccupied compositional variable, and given that each occupied part is coupled with a sentence of the type 'this compositional character is to be taken as an x in mathematical terms and an x* in biological terms'. Not so, however, in the case of Fig. 13, Fig. 16, Fig. 25 [a], [b], Fig. 26 [a], [b], [c] and partially also Fig. 21, Fig. 22, Fig. 27. The suggested construal of these figures makes use of every compositional variable within their vaguely framed corpuses, and promotes, due to a lack of clear limits in the functional occupation of the geometrical, topological and graphically describable composites, the use of vague descriptors, such as vague locators, like "area" (Falk et al. 2009, p. 171), or vague quantifiers, like "about" (cf. Schittler et al. 2010, p. 7), which keeps the extraction and defence of interpretative derivatives rather general and tendency-orientated (cf. e.g. "In Fig. 6 [here: Fig. 16] it can be seen that the percentages of cells depend on the amounts of stimuli", Schittler et al. 2010, p. 7).

However, vagueness does not only correlate with the functional occupation of every compositional component (which is well justified in case of Fig. 16, since each coloured point corresponds to one solution of the model, altogether too many to be discerned and memorised individually). Vagueness can also occur in some sort of inverse situation: Fig. 15, for instance, draws on a colour code; Fig. 14, however, is kept in black-and-white. While this difference between the two figures is fairly easy to discern, there is no explicit reason for it; by contrast, Fig. 13 and Fig. 16 offered a hardly tractable multitude of instances, but clear codes for each. One possible explanation for the compositional incongruence between Fig. 14 and Fig. 15 is that Fig. 14 shall not be mistaken for a key to Fig. 15 (et v.v.); in other words, the compositional differences might be used to point out that the two figures are linked to different sources of information. However, as said before, this purpose is not confirmed, which leaves one to assume that it is either part of an implicit, community-specific knowledge about the figures, or there are, indeed, compositional differences, which just do not have any particular function. All the same, it remains to say that one expects conceptual accounts of these domains to offer room (and perhaps keys) for making sense of the differences in the compositional profiling and complexity.

3.3.5 Observations on the prevalence of figures

One last remark shall be devoted to the amount of figures in the consulted papers, which offers another, though merely quantitative, option for approaching the status of figures. It is striking in this context that both papers (Schittler et al. 2010 and Falk et al. 2009) make references to one or more figures in almost every section. Only the introduction and the conclusion are left blank in the first paper (Schittler et al. 2010). By contrast, the second paper offers an introductory preview (cf. Fig. 19) and further cross-sectional references to the figures in the whole body of the paper. The sheer prevalence of figures (as structural components of the inspected paper and as invoked bases of information) thus strengthens the impression that they are deeply connected to both the research document as well as the documented research. This gives at least psychological support to the premise about the integral role of figures in these documents (as perhaps in others; cf. Downes 2012, p. 116ff.), which will now be subjected to conceptual review and refinement.

IV Epistemic Weight and Epistemic Surpluses

It now remains to devise a framework against which it is possible to judge statements on the epistemic weight and surpluses of pictorial means. To this end, I will first scrutinise the literature for accounts that allow for the occurrences of pictorial means in science as opposed to linguistic means (cf. 4.1). The key finding of this survey will be that most of the current approaches tackle only one or two concepts of a four-part conceptual module. The latter is however needed to guarantee an efficient framework (cf. 4.1.4). The account that comes closest to the expectations in this and other regards is the one by L. Perini (2002ff.). I therefore devote an in-depth analysis to her view, particularly on the epistemic weight of pictorial symbol systems (cf. 4.2.1), and suggest modifications concerning the conceptual architecture (cf. 4.2.2). These modifications – especially the introduction of *ad hoc syntaxes* – should then also allow for practices that could not be taken into account so far, such as *microscopy-like* images (cf. 3.2). In terms of the epistemic surpluses, I will offer a more independent proposal for defining an open list of specific capacities (cf. 4.3). The limits of the resulting positions and the supportive conceptual apparatus will finally be revealed (cf. 4.4), so that future contributors can decide to search for other approaches – also depending on the developments in the academic fields.

4.1 Recent accounts of picture-like means in science

Although things have improved, it is still to be noted that "[m]ost people writing about visual representations attempt to explain their roles in science without including an analysis of their symbolic features" (Perini 2002, p. 33). In other words, we find many explanations with reference to reported purposes, but we lack functional explanations that rely on a synopsis of functional programs and the contrasted means (cf. 2.2.1.2). In order to account for all factors, I will devote an in-depth analysis to the only account of which I think is a source and valuable example of such a synopsis (cf. 4.2). But before this, I will work my way through the research landscape and confront the previous accounts with arguments on their significance and soundness. The significance shall depend on their compliance with the gathered expectations and demands on a conceptual view (cf. 3.3). Soundness shall be defined as the lack or rarity of counter-examples and -intuitions.

4.1.1 Operational accounts: analysing modes of implementation

One option of analysing the notion of picture-likeness is to expound the *procedures* that are essentially linked with it; this may also help establish a link to scientific practice.

This option seems to be convincing, especially when bearing in mind that computational scientists conceive of visual means in terms of multi-stage procedures rather than finished products in little frames. This is indicated by the researcher's self-conceptions being launched in the targeted fields. Visualisation is accordingly "more than a tool; it is a way of thinking" (Wong & Bergeron 1997, p. 4). The associated aim is therefore to explore "key trends and relationships among the variates [of a data set]" (Wong & Bergeron 1997, p. 7). In accordance with this agenda, originating from computational science, the leading paradigm of scientific visualisation is a *sequence of procedures*, which is referred to as the "visualisation pipeline" (cf. Fig. 32; Ertl 2009, 2nd lecture).

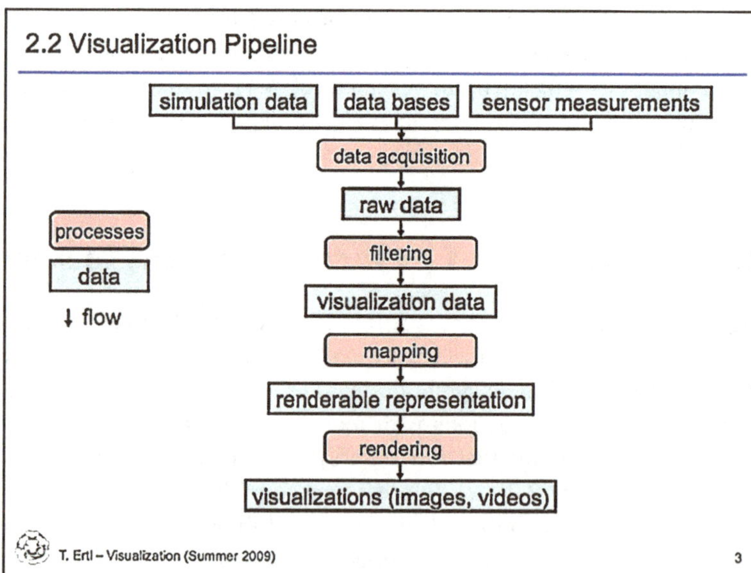

Fig. 32: Visualization pipeline (cf. Ertl 2009, 2nd lecture)

The process commences with the acquisition of data from simulations, measurements or other sources. In the next step – the so-called 'filtering' – the data is prepared for its transcription into a visual code. This is accomplished by various means, such as the removal of irrelevant data and the extrapolation to missing data. Data sets that have been pre-processed in this way are then transcribed – which is called 'mapping' – into a geometrical or topological code. The features of the data set (esp. values and relations) are preserved as best can be. In the last step, the mapped data is 'rendered'. That is to say, it is endowed with graphical cues, which are chosen in consideration of the user's skills and purposes. For example, the rendition can be made more 'realistic' (Ertl 2009, 2nd lecture) by adjusting it to a prior (e.g. biological) version.

As we can expect, the last step of this process – the 'rendering' – attracts the most attention in philosophy. This might be due to some sort of classical education or background in light of which this is seen as a process of creation, which brings about an *oeuvre* of potential representational and methodological value. However, this analysis of the 'visualisation pipeline' (which is itself an analysis of visualisation practices) would be ignorant of the fact that the process is not finished at a fixed point; there can be pre-versions or later revisions, which are redundant for the pipeline model but crucial in practice. Moreover, the process is not irrelevant for the resulting rendition, for the simple reason that it is part of the process of defining its own (perhaps only interim) end. Last but not least, one can hardly maintain that only the interim result instantiates the idea of a visual means. That this cannot be the case becomes clear when considering Fig. 13, for example, which is quite different from the published version (Fig. 14), but which is still an instance of a picture-like means.

For these reasons, any attempt to abstract distinct kinds of entities from such a series of procedures is faced with the problem of trapping them at some point in the workflow. The latter indeed defies the category of a 'piece of work' (which is drawn from aesthetics).

In this section, I will discuss an alternative conception of visual means – one that refrains from further abstractions and takes a series of operations as being the concept or idea of the entity under consideration. In other words: it seems perfectly possible to present a list of operations in response to the question, *What do we refer to as a visual means?* One may likewise define the idea of a 'house' or 'house-like means' by saying that it is built in such and such ways (e.g. endowed with a basement, rooms and a roof) and then adjusted to individual ways of living in such and such ways, etc.

In order to evaluate this strategy of conceptualisation, I would like to discuss the pros and cons, using P. Bridgman's essay *The Logic of Modern Physics* (1927) as an example. Bridgman has presented the first and certainly an "extreme" (Chang 2009) manifesto in support of operation-based analyses. For this reason, his essay may contribute to the discussion, or, rather to the recapitulation of crucial objections and provisos against this type of approach.

Bridgman's goal is to get rid of ill-substantiated claims about characteristic properties, which according to him have corrupted the key concepts in physics, such as the concept of time[53] (cf. Bridgman 1927, p. 4). Along the lines of a meta-methodology, Bridgman then proposes a range of amendments to improve concept-building in physics. In the eyes of the critics (cf. e.g. Chang 2009), Bridgman calls for getting rid of properties *on principle* when it comes to forming a concept. In terms of the resulting void of information, he argues that the latter is only lamented for psychological reasons, and that properties are indeed dispensable since "we mean by any concept nothing more than a set of operations" (Bridgman 1927, p. 5).

He uses the concept of length as an example. It is analysed as follows:

> "We start with a measuring rod, lay it on the object so that one of its ends coincides with one end of the object, mark on the object the other end of the rod, then move the rod along in a straight line extension of its previous position until the first end coincides with the previous position of the other end, repeat this process as often as we can, and call the length the total number of times the rod was applied" (Bridgman 1927, p. 7).

Bridgman is clear about the limits of this analysis. It does not, for example, apply to the length of distances between *moving* bodies, since this requires a different kind of measurement technique (cf. Bridgman 1927, p. 7f.). Moreover, according to his agenda, the analysis is only valid if every single concept referred to in the analysis allows for such an analysis, too. However, before we sort out the drawbacks, let us briefly recall the merits of such an analysis with special regards to the concept of picture-likeness.

An operational analysis indeed draws attention to the implementation of picture-like entities. This helps integrate the intuition about the practical implications of picture-likeness (cf. 3.3.2). Moreover, it fits hermeneutic theories, especially

---

53 In order to give an example, Bridgman mentions the claim that absolute time flows equably (cf. Bridgman 1927, p. 4). This has long been considered as the adequate view of time, but it turned out to be false against the background of Einstein's special theory of relativity. The rejection of this belief was experienced as a "shock" (Chang 2009), and it made Bridgman contemplate the options of avoiding such failures. As mentioned above, he suggests an operation-based technique of conceptualisation (cf. Chang 2009).

those that conceive of interpretation as a *process* of co-creation (cf. e.g. Iser 1975). Besides, an operational analysis could pave the way for modelling functional programs, which in turn are needed to discuss arguments about the required capacities of the involved means (cf. 2.2.1.2).

Keeping these advantages in mind, it is time to face the objections against Bridgman's account. I will present only a selection of them, which suffices to evaluate the account.

From a general epistemological point of view, to begin with, it can be objected that "[o]perational definitions are not required for all useful concepts" (Chang 2009). The concept of speed, for instance, has not much suffered from the shift to Einstein's paradigm. Therefore, the initial crisis – the failure of conceptualisation – does not justify a *general* overhaul of conceptualisation, as it concerns only *some* concepts.

More specifically, it does also not justify the restriction to operational accounts, because the concepts of failure and adequacy do not *themselves* correlate with invariant sequences of operations. Different concepts can fail or succeed for different reasons; there is no one defining way of determining adequacy or failure in this sense. Still, Bridgman uses the categories of adequacy and failure, which means that he relies on an exception to his own theory. Bridgman thus fails to meet his own demands, because the targeted idea of adequate concepts cannot be analysed in the proposed adequate way.

After this brief reflection, we can thus agree that Bridgman throws the baby out with the bathwater, meaning that his agenda turns out to be "overly restrictive" (Chang 2009). An operation-focused analysis could still work in *some* cases though.[54] Therefore, it is worth mentioning another problem pertaining to its specific potential.

A further problem arises when attempting to adjust the operational analysis to a basic methodological intuition, namely the intuition that a concept is adequate

---

54 Some consider it to be "highly doubtful that Bridgman intended to advance a precise and universal theory of meaning" (Chang 2009). Bridgman himself was reportedly puzzled about the fact that his account obtained the status of an '*-ism*' ("operationalism" Chang 2009). The problem is that his essay supports both interpretations. On the one hand, he asserts: "A precise analysis of our conceptual structure has never been attempted […]. It will never be possible to give a clean-cut logical analysis of the conceptual situation" (Bridgman 1927, p. 13). On the other hand, he launches the following prospect: "Let anyone examine in operational terms any popular present day discussion of religious or moral questions to realize the magnitude of the reformation awaiting us", Bridgman 1927, p. 17).

if it is distinct enough to support casuistry. From the viewpoint of an operational analysis, this means that "[w]e must demand that the set of operations equivalent to any concept may be a unique set, for otherwise there are possibilities of ambiguity in practical applications which we cannot admit" (Bridgman 1927, p. 5). Consequently, "we must recognise in principle that in changing the operations, we have really changed the concept" (Bridgman 1927, p. 13). This being the case, the obvious problem is to cope with the variant and variable operations that are linked with *one and the same* concept; because, needless to say, the operations that are associated with the attribute 'pictorial' cannot straightforwardly be reduced, for instance to those that belong to the 'visualisation pipeline'. Bridgman in turn objects that this semantic variance "is dictated only by considerations of convenience" (Bridgman 1927, p. 13); and conventions are not reliable as guides to adequate concepts. Still, and even more as he admits that, "in the transition regions where nature is getting simpler [...], a certain haziness is inevitable" (Bridgman 1927, p. 13), the question very much is how one should distinguish between concepts that are distinct enough and those that are not. It would seem that, in order to answer this question, an operational analysis of the concept of operational analysis is needed. Yet, as we already learned, this is doomed to infinite iterations, since it would also require an analysis of the analysis of the analysis, etc. A further, more specific objection against 'operationalism' thus is: either we accept a "fragmentation" (Chang 2009) of the landscape of conceptions, since all differences in practice would amount to differences in conception, which leads to a *map* of practices rather than to a *taxonomy* thereof; or we get stuck in an infinite regress, since we would be forced to operationalize notions that are pre-supposed in the idea of operationalization (e.g. the idea of *analysis* or *adequacy*). Both options are inconvenient.

As can be expected, a pure operation-based approach also reveals deficits in comparison to other options of conceptualisation. Firstly, since one cannot insist that operational analyses are always the better choice, they compete with property- or function-based accounts (cf. 4.1.2; 4.1.3); the latter have a good chance of outpacing their operation-obsessed rival by providing class-defining conditions that generalise upon practical varieties. Secondly, the steady pressure to establish new concepts in order to account for the diversity of practice collides with the intuition that one and the same concept can profit from revision; it may thus be improved without changing entirely.

Despite the compelling criticism, Bridgman's essay is nevertheless worthy of note, since "[w]e can take operationalism as a useful and practicable caution not to make conceptual extensions without operational grounds" (Chang 2009). We may not have found a methodological role model on that score, be we have found further support for the involvement of practical correctives (cf. 3.1, 3.2) and implications (cf. 3.3.2).

4.1.2 Ontological accounts: analysing traits of occurrence or identification

While an operation-focused analysis of the characterising traits of *picture-like* instances scores exceptionally well when it comes to absorbing information about practical implementations of the *definiendum* (cf. 4.1.1), it falls behind when it comes to justifying and limiting the absorption of practical information. The lesson to be learned from this is that, besides appreciating cases and examples from practice, it is no less important to suggest what they might have in common, for only such acts of abstraction can actually be tested as to whether they guide or confuse casuistry. For this (perhaps trivial) reason alone, it should be welcomed if a conceptual analysis yields a *tertium comparationis* – in terms of a common idea or distinguishing mark – that aids the sub-grouping of groupings, such as of those listed by M. Lynch and S. Woolgar (1990): "graphs, diagrams, equations, models, photographs, instrumental inscriptions, written reports, computer programs, laboratory conversations, and hybrid forms [...]" (p. 1).[55] To stress this point again: only if one tries further to discriminate between or generalise upon such lists of instances, will *problems* of conceptualisation arise, like the question, whether it makes sense to distinguish between different kinds of visual means, or whether the sub-kinds can be seen as integral to scientific practice, etc. Accordingly, those who acknowledge these questions as veritable problems must refer to concepts or conceptual frameworks that do not just *map* practices (as suggested above; cf. 4.1.1) but that *classify* them.

Switching back from the methodological to the conceptual discussion, I will now start with surveying those approaches that rest on the idea that sets of properties can be meaningfully and profitably comprised under a common concept. However, it will turn out that, especially in the early days of the debate, the

55 M. Lynch and S. Woolgar refrain from introducing more general distinctions, arguing that the listed items are *family-like* (cf. Lynch & Woolgar 1990, p. 1f.; cf. 2.1.1). Yet this makes it difficult for them to adopt the motivators of the debate, particularly the dichotomy between pictorial and verbal means, which is defused rather than expounded, when stressing their common origins in the first place.

accounts leave room for improvements concerning their conceptual infrastructure and explicitness. The following survey will thus be appreciated as a reason for continuing rather than for closing the debate. After the survey, which is followed by another survey with a slightly different focus (cf. 4.1.3), the upshots will be discussed in more detail (cf. 4.1.4).

In this context, it has become obligatory to mention Ch. S. Peirce, especially in research on diagrams (cf. Schmidt-Burkhardt 2009, p. 165). Peirce is known among other views for his contribution to semiotics, which is often condensed into the following distinction: "A sign is either an icon, an index or a symbol" (Peirce 1955, p. 104). The related conceptual footing triggered an exegetical debate in its own right. One reading of the fragments says that the distinction is "justified with regard to the specific relationships that signs can have to their referents. In highly simplified terms, one may say that an icon establishes a relation based on similarity or analogy, an index arises from a certain mode or within a certain context of establishment, and a symbol is tied to its referent thanks to conventions or patterns of thinking. [Moreover,] due to the structure-based similarity between sign and referent, Peirce subsumed the diagram into the class of icons, and thus to the class of images in the broadest sense"[56] (Schmidt-Burkhardt 2009, p. 166).

Unfortunately, the extensive comments on Peirce's taxonomy cannot be reviewed here in greater variance and detail. The critical point in the context of this thesis is that Peirce is also debated in comparison to other initiators of the debate, such as N. Goodman (1976), who surpasses Peirce's account regarding the current gusto in the philosophy of science (a similar situation as in history of arts, where Peirce is rarely applied; cf. Wyss 2006, p. 46).

N. Goodman differentiated between pictorial and diagrammatic (resp. graphical) symbol systems, which should be approached as *systems* (and not as isolatable units of signs and referents, as Peirce suggests) and *all* taken as established on *conventional* bases. Apart from rejecting similarity as a constituent of representations thanks to which Goodman's view is adaptable to recent accounts of scientific representation, he also suggests that "[t]here is a syntactic difference between

56 Translation of the following quote: „[Die Einteilung] wird durch das spezifische Verhältnis des jeweiligen Zeichens zu seinem Referenten begründet. Stark vereinfacht beruht die Objektrelation beim Ikon auf einer Ähnlichkeit oder Analogie, beim Index auf einem Modus oder Kontext und beim Symbol auf einer Konvention oder Denkgewohnheit. Das Diagramm ordnete Peirce aufgrund seiner strukturell-abstrakten Ähnlichkeit mit dem Referenzobjekt den ikonischen Zeichen und damit im weitesten Sinne der Gruppe der Bilder zu" (Schmidt-Burkhardt 2009, p. 166).

pictorial and diagrammatic systems"[57] (Schmidt-Burkhardt 2009, p. 166). From this and other differences, further sub-variants of the systems can be derived, matched up with the procedural traits drawn from practice and finally examined on their roles in function-ascribing programs. However, as far as I can see, this has only been tried by L. Perini (2002ff.) and, briefly, by Cummins (1975). Yet before turning to the preferred reference views, I shall withhold the supporting arguments for now and continue with the review – now already in defence of *this* particular choice (esp. Perini 2002ff.; Goodman 1976).

In fact, there is an earlier, almost-forgotten advocate of the notion that visual means are established on conventional bases *only*. Ludwik Fleck noted, quite independently it seems, that, "[i]n order to see something, one needs to know what is relevant and what not; one needs to be able to discern the background parts of the image; one needs to understand about the category of the [foreground?] object"[58] (Fleck 1983, p. 148). Fleck adds an analysis of calligraphies in support of the intuition that the relation between compositional and representational modules must be *pre-defined*, in order for a viewer to be able to approach a set of "*Gestalten*" (Fleck 1983, p. 153) as possible instances of a certain coding system. In other words: "Knowing about signs opens up the possibility of recognising them"[59] (Fleck 1983, p. 153). Hence, although Fleck does not exhaust the full depth of a semiotic analysis – for he does not provide any means of distinguishing between different types of coding systems – he nevertheless captures a basic intuition, namely that all sorts of signs are *means*. In this context, he also unveils the consequences that this has for their mastery, such as the need for tutorials (cf. 2.2.2.1; III intro).

While Fleck's account remains a little isolated in the history of the discussion due to his late rediscovery by Kuhn (1998), M. Rudwick acted more programmatically, when he stated "a lack of any strong intellectual tradition in which visual modes of communication are accepted as essential for the historical analysis and [for the] understanding of scientific knowledge" (Rudwick 1976, p. 149).

---

57 Translation of the quote: „Der Unterschied zwischen einem piktorialen und einem diagrammatischen Schema ist syntaktischer Natur" (Schmidt-Burkhardt 2009, p. 166).

58 Translation of the following quote [square brackets added] from the German edition (which in turn has been translated from Polish by Th. Schnelle): „[u]m zu sehen, muß man wissen, was wesentlich und was unwesentlich ist, muß man den Hintergrund vom Bild unterscheiden können, muß man darüber orientiert sein zu was für einer Kategorie der Gegenstand [im Vordergrund?] gehört" (Fleck 1983, p. 148).

59 Translation of the following quote from the German edition: „Die Kenntnis einer Gestalt, schafft die Disposition, sie wahrzunehmen" (Fleck 1983, p. 152).

This concerns the sciences as well as the sciences that *deal* with science: "In the hierarchy of our educational institutions, visual thinking is simply not valued as highly as verbal or mathematical dexterity; and an academic discipline such as the history of science has inevitably been infected by this bias of educational values" (Rudwick 1976, p. 150; cf. 2.3). Rudwick picks up this bias and tries to tackle it. To this end, he examines what he refers to as the "visual language" (Rudwick 1976) used in early geological science. The way he chooses terms makes apparent that he aims to rehabilitate visual means by bringing them close to linguistic means, which is in fact the inverse strategy to highlighting the visual qualities of (esp. written) language (cf. Krämer & Totzke 2012). In the eyes of his critics, what is particularly pioneering about move is that, "the form of visual representation is related to their function" (Perini 2002, p. 20). However, a closer look reveals that this mainly concerns the *pre*conditions of *manufacturing*, such as the costs and tractability of engraving and printing equipment. At least, this is suggested when Rudwick points out, for instance, that "copper engravings were excellent for line drawings and maps requiring fine detail" (Rudwick 1976, p. 155), while a "disadvantage of engravings, as a potential medium for visual communication in science, was that they had to be printed separately from the text and generally on a different kind of paper" (Rudwick 1976, p. 155). Along with these findings, Rudwick offers a range of concepts. However, he struggles with balancing more general and more specific scopes of systematic integration, as with his definition of maps:

> "A modern geological map is an attempt to depict on a two-dimensional surface what cannot in reality be seen at all except in isolated exposures, namely the outcrops of rocks which are generally concealed by soil and vegetation" (Rudwick 1976, p. 159).

This might work for geological maps – although one misses a hint to the relevance of rock types – but not for all maps used by geologists, not to mention other types of visual means in geology. Hence, even though the agenda is promising, there is more work to do – also in regard to the relation between different visual modes. Rudwick was well aware of this:

> "I have deliberately refrained in this article from attempting to analyze the relation between visual and verbal forms of communication in this (or any) science. The historical study of visual communication in science is still such a neglected area of research that I have thought it better to confine this article to an outline description and tentative interpretation of the visual language of geology, considered in its own terms" (Rudwick 1976, p. 181f.).

Rudwick, as it turns out, furthered the idea of conceptual research a little more than the agenda of conceptualisation itself. This changed in the following years. In

1988, M. Lynch presents a study on how “visual displays” (Lynch 1988, p. 201) contribute to the “constitution of the scientific object” (Lynch 1988, p. 201). His claim is that visual means are “essential to how scientific objects and orderly relationships are revealed and made analyzable [...] [even more as] visual documents are used at all stages of scientific research” (Lynch 1988, p. 202). In search of support for and substantiation of this claim, Lynch goes through a selection of visual items, asking not only *why* they are used, but also how it is that they *can* be used this way (thus paving the ground for a functional analysis; cf. 2.2.1.2). He chooses the process of *selection* for one of the two close-up inspections (cf. Lynch 1988, p. 203ff.). In Lynch’s terms, selection is a regular mode of coping with experiences: it covers the acts of picking out, stripping away and transforming features. These acts, which the mind seems to foster anyway, would be aided by visual means in the realms of science. The visual aids in turn, so the argument goes, have a direct impact on their own use – not only in terms of enabling it, but also in terms of shaping its consequences. A transformation of features, for instance, happens more or less intentionally when one “strives to identify *in* the particular specimen under study ‘universal’ properties which ‘solidify’ the object in reference to the current state of the discipline” (Lynch 1988, p. 205). In other words, the claim is that the selection of relevant features is guided by a more general, multi-modally established ‘image’ of the research object, which in turn is strengthened as a role model by each analysis that it is capable of guiding. This claim is also substantiated by the second close-up inspection, which is devoted to the process of *mathematisation* (Lynch 1988, p. 217). The main point is here basically that visual means pre-configure the way a research object or system is represented in mathematical terms – a claim that also seems valid against the examples above (cf. esp. 3.1.5, 3.1.6). Most importantly perhaps, Lynch performs a selection himself, in the way that he reveals notable traits of the studied visual composites. The latter would be organised in a way that is “discrete, outlined, and defined by relations of separation and overlap” (Lynch 1988, p. 225). He assumes that those traits are the hinges, which allow for the adjustment of the entire functional compound to the “disciplinary eidos” (Lynch 1988, p. 216), that is, to the prioritised renderings that offer criteria for being ‘realistic’. This also sounds familiar after the inspection of the microscopy-like images (cf. 3.2.9; 4.1.1). In fact, the thing one misses most about Lynch’s approach is an extension of the agenda, especially since he does not indicate why selection and mathematisation should be the *only* procedures that are essentially aided by visual means. Also, it would be interesting to learn whether the analysis of visual items (e.g. as being discretely composed) follows a “disciplinary eidos” (Lynch 1988, p. 216) as well. Since, remaining blind to its own conditions of research – which is conducted *with* as much as *on* – visual means is a notorious trap. In fact, from the authors surveyed, only Fleck (see above) and Giaquinto (see below) draw attention to the fact that the analysis of compositional traits is not trivial in and of itself. Still, the good impression remains: Lynch captures valuable intuitions in methodological as well as conceptual terms. Actually, he may even serve as a root to the notion about the entanglement of justificatory frameworks, against which visual means can obtain or be exempted from different roles: "The mathematical/natural/literal unity of particular displays can be taken apart. Data appearing as graphic inscriptions can be treated as detachable signs, independent of the object to which they refer" (Lynch 1988, p. 228).

Roughly fifteen years after Rudwick pointed out the lack of research on visual means, W. Wimsatt notes increasing interest among "scientists in all fields who must deal with problems of how to present large quantities of complex multidimensional data in an intelligible fashion" (Wimsatt 1990, p. 111). He considers new empirical studies on the usability of visual aids, also by non-tutored users, who would have proven capable of discerning compositional features such as boundaries and connectivity (cf. Wimsatt 1990, p. 116). This is set in correlation to a list of scientific tasks, which "make visual or spatial representation particularly appropriate: (1) analysis of highly multidimensional data […], (2) The need to see, or to find, in such data (a) similarities in diverse cases in spite of their differences, or (b) differences in spite of their similarities, or (c) to factor perceived global dissimilarities into localizable similarities and differences. (3) The need to look for relationships among patterns on several different size scales or levels of organization […]. (4) The need to combine or use jointly information from different perspectives on (i.e., different ways of accessing or measuring) the same object […], or (5) to determine where there are boundaries between objects, or to identify whether there is a common object, property or cause, behind apparently correlated patterns" (Wimsatt 1990, p. 117). The claim about the specific appropriateness of visual means is justified with respect to their composite profile, which would offer enough variables for the transcription of "multidimensional data" (Wimsatt 1990, p. 116). From the scope of usage, Wimsatt also derives that, "[t]o the extent that theories aim to organise and simplify […], pictures can have a theory-like status" (Wimsatt 1990, p. 112). Unfortunately, even though Wimsatt offers the best analysis of the notion of *data exploration* so far, his epistemological analysis is lacking in rigor – with regard to the claimed capacities, which are often confused with the tasks, as well as with regard to the ontological and epistemic peculiarities of *non*-pictorial aids, which are not explicitly set in contrast to the findings.

R. Giere (1996) took note of the previous works in several respects: on the one hand, he offered an explanation for the bias against visual means (at least pertaining to the 20th century) by calling it a legacy of logical empiricism, which excluded any "fundamental role [...] for non-linguistic entities like pictures and diagrams" (Giere 1996, p. 270; cf. 2.3.2). On the other hand, Giere began to reflect on the deficits of prior attempts of rehabilitation: "Significantly, relativist sociologists of science were among the first to investigate the role of pictures, diagrams, and other non-propositional forms of representation in science" (Giere 1996, p. 271); however, to the extent that they aim at comprehensive histories of the spatiotemporally encircled practices in the first place, "they deny any genuine role for visual images in science" (Giere 1996, p. 272). In other words, if the aim just is to tell a detailed history about the work of scientists, the neglect of visual means is not controversial, it is a clear *mistake*. Yet, if the aim is to refine the related notions (e.g. about entities bearing epistemic weight), the neglect of visual means is a matter of debating and matching the currently proposed frameworks. And so, having the logical empiricists on the one side, who bought into a linguistically biased set of premises, and the relativist sociologists on the other side, who concentrated on the common grounds of pictorial and linguistic procedures, Giere opted for a "middle way" (Giere 1996, p. 272). This 'way' shall combine the virtues of tackling *basics* while staying *context-sensitive*, meaning that it shall allow for measures formerly reserved for language(s), such as inferences and judgment, without over-interpreting practices. Unfortunately, this promising agenda turns into a study of methodological involvements rather than of the involved means. The consequence is that statements such as, "[o]ne cannot argue, I think, that the images are *logically* essential" (Giere 1996, p. 284), remain without appeal to the mean's potential. It is also not excluded that images can be part of procedures, which rely *also* on logic.

After all, and this is the impression left by the first phase of the discussion, the project of framing claims about the epistemic weight and uniqueness of visual modes would profit from tackling *four* notions, not just two: in particular, one will demand an account of both the compared functional programs (i.e. the tasks to accomplish) *and* the employed means, since one cannot avoid referring to the tasks *and* the tool's capacities to accomplish them when explaining that the tools can be tools in this sense (cf. 2.2.1.2). However, this apparently requires the clarification of four variables, which, considering the options of combination, endows this task with the dimensions of a small research program (i.e. the second phase of the debate, which gives rise to monographies [cf. 2.3.3], journals [e.g. *Rheinsprung 11*] and series [e.g. *Bildwelten des Wissens*]).

In the present discussion (counted from the year 2000), the amount of contributions has increased rapidly. Consequently, it is difficult to condense all the inputs into clear-cut positions and representative examples. In general respects, one may say that some of the frequently repeated assertions have become truisms in the meantime.

R. Kötter, for example, has reinforced the arguments against the 'naïve' view of visual or other model-based representations, according to which images display states of affairs (cf. Kötter 2007, p. 163; Wedekind et al. 1998, p. 266f.). As Fleck has pointed out, this view fails in the face of the personal and social requirements and related variances of visual literacy. Kötter adds in support of this stance: "in order to understand such pictures [i.e. a morphological plan or geological map], it will not suffice to compare them with the so-called 'reality' – rather, one needs to understand the text to which they are assigned"[60] (Kötter 2007, p. 163). That is to say that it is part of the very nature of visual means (at least in the inspected scientific realms) that they be related to another source of information and thus to another representational system or instance thereof (cf. 3.3.1). Kötter's input, in this context, is a taxonomy of "basic functions" ("Grundfunktionen", Kötter 2007), which is in fact an overview of methodologically justified variances of this particular relation. The corresponding classifiers have already proven useful in an earlier chapter (cf. 3.3.3). Thus, in the course of a depiction ("Abbildung", Kötter 2007, p. 166ff.), for example, a set of characteristic features of a research object is represented in accordance with descriptions of their relative locations, proportions and shapes. The composite result can either be used for the purpose of illustration (i.e. as an aid to summarising, understanding or memorising) or documentation (i.e. as a testimony; cf. Kötter 2007, p. 166ff.). He further discerns between visualisation ("Verbildlichung", Kötter 2007, p. 171ff.), that is, the transcription of spatially defined features into a topological, geometrical or graphical code, and exemplification ("Veranschaulichung", Kötter 2007, p. 176f.), that is, the explication of core aspects or ideas of a concept, model or theory by the aid of an analogous representation (Waddington's 'epigenetic landscape' should be an example of this type; cf. Fig. 33). Now, while these categories seem to encapsulate some practical traits, the problem with Kötter's approach is that he does not put the typified measures into relation

60 Translation of the quote: „um solche Bilder [einen morphologischen Bauplan oder eine geologische Karte] verstehen zu können, genügt es nicht, sie mit der so genannten „Wirklichkeit" zu vergleichen, man muss vielmehr den Text verstanden haben, dem diese Bilder zugeordnet sind" (Kötter 2007, p. 163).

with his notion of images (discussed on an earlier occasion; cf. 2.1.2.4). This is neither wrong nor inconsequential, but, according to the four-part research agenda revealed above, it leaves promising options unused that would perhaps shed more light on the final statement, which says that images often have an indispensable heuristic function (cf. Kötter 2007, p. 177).

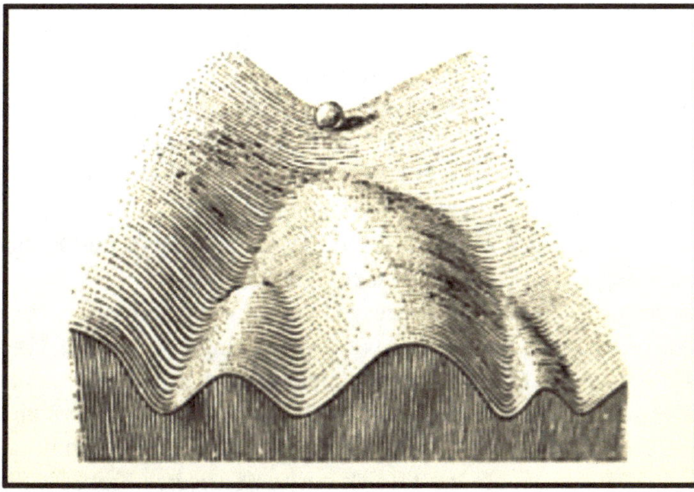

Fig. 33: C. H. Waddington: Intra-cellular decision-making (cf. Waddington 1957, p. 29)

In the same year of publication, M. Giaquinto (2007) tackles some of the afore-noted gaps. He also goes deeper into the consequences of the central claims, using examples from mathematics and the cognitive sciences. For example, he demonstrates why it is already difficult to pinpoint the *compositional* profile of a graphical system. While the foregoing works focused on the relation between graphical composites and their various functions in the first place, Giaquinto makes us aware of an earlier level of problems: He elucidates that, even if one is well-versed with geometrical, topological and other coding schemes, a compositional analysis can cause confusion and disagreement due to phenomena such as the emergence of shapes or the irresolvable vagueness of properties. As he puts it more concretely, "[v]isually perceiving edges or borders of surfaces [...] does not necessarily involve seeing lines that mark those. [...] Perceptual borders can be constructed from contrasts in the luminance of adjacent regions, or from lines, or by the more indirect means involved in perceiving illusionary borders [...]" (Giaquinto 2010, p. 14). What is more, one and the same optical excitement can be assimilated differently under the impact of alternative categories. For example, many people do (but would not be forced to) recognise a square, even when confronted with a set of *nearly* closed lines; and they would recognise a diamond as soon as the square is turned for *about* 45°. For such phenomena, Giaquinto assumes that "[t] he degree of imperfection allowable is not something one can specify" (Giaquinto 2010, p. 27; cf. also Fleck 1983). The crucial consequence is that even on the syntactical (resp. compositional) level, acts of interpretation are required, too. A naïve view is thus to be rejected here as well. Still, this is not the end but just another benchmark of the epistemological agenda. Giaquinto offers a range of widely accepted proofs and their establishing frameworks that suffice to challenge the view that "the utility of visual thinking in mathematics is only psychological, not epistemological" (Giaquinto 2007, p. 1). Tackling this view distinguishes him (and also Perini 2002ff.) from previous accounts, which mainly contribute to the notion that visual means are auxiliaries to the assimilation of information, which was even true for the logical positivists (cf. 2.3.2, 2.2.3).

The last series of contributions surveyed in this section originates from the history of arts, which takes an essential part in exploring the status of visual means in scientific areas insofar as it provides a methodology and related tradition of research for representing, contextualising and inventorying picture-like and picture-based practices (cf. 2.2.1.3; III intro; IV intro). Of the many contributions (see also 4.1.3, 2.3.3), it seems fitting to pick out some excerpts from H. Bredekamp (esp. Bredekamp 2000, 2008, 2010), who has supplied the respective area of research with several reference analyses and concepts. Moreover, his contributions to conceptualisation reveal a peculiar mixture of strategies, which could be a consequence of the fact that he has accompanied the discussion all over the second phase and thus took the opportunity to refine and readjust his account. On that note, his work can be seen as the missing key in this survey to the revelation of research *tendencies*, which have not yet been analysed for the whole field (for action-theoretic accounts, cf. Seja 2009). As is often the case (cf. 2.1.2.1), Bredekamp's classification of pictures *as* pictures follows for the most part a broad, case-based and implicit strategy, which allows for "any kind of creation"[61] (Bredekamp 2010, p. 34). This involves all sorts of things, in fact all that show a "minimum of human treatment"[62] (Bredekamp 2010, p. 34) – the concomitant extension of the research

61 Translation of the following quote: „jedwede Form der Gestaltung" (Bredekamp 2010, p. 34).

62 Translation of the quote: „ein Minimum an menschlicher Bearbeitung" (Bredekamp 2010, p. 34).

program is a welcomed effect and has arguably contributed to making people aware of visual items outside the game of art, although it excludes, for instance, the range of so-called 'natural' images (i.e. images occurring in clouds, silhouettes of mountains, etc.; cf. Mitchell 1984, p. 504ff.). In terms of the more restrictive accounts, Bredekamp seems to appreciate the work of G. Boehm (1994), who argues that images are defined (at least marked) by the so-called 'iconic difference' ("*ikonische Differenz*", Boehm 1994, p. 30). This phrase captures the notion that entities are (to be) addressed as images only if they are marked by a certain *contrast* between their composites, which have a representational function in their own right, and the entire compositional and representational unit, which somehow emerges from the arrangement and interplay of its parts. I must admit, however, that I find this oft-referenced account quite difficult to grasp. If it says that pictures are decoded step by step, yet always in relation to prior and later readings, the conception leaves only a few options for cashing out a categorical difference to non-pictorial symbol systems; in fact, it melts with the notion of *systems* in general. J. Steinbrenner (2009) also objects, "[n]obody doubts that the components of pictures contribute to their meaning and that they can have an individual meaning as parts. What is missing rather are commonly accepted methods for determining these components"[63] (Steinbrenner 2009, p. 285). Bredekamp seems to be aware of the critique, as he shows sympathy for other strategies of conceptualisation, too. For instance, he also opts for the criterion of family-likeness (cf. 2.1.1), when he asserts that "pictures can only be recognised as such when consulting reference series"[64] (Bredekamp 2010, p. 29). Yet this leaves the distinction between pictorial procedures and other domains of human treatment still unaccounted for, which is presumably why he merges a third attempt of conceptualisation with the afore-noted ones. In particular, he adds together what I distinguished as 'picture-like' and 'picture-based' procedures (for justification see 3.3.1). As far as I can see, he conceives of *both* in terms of "picture-acts" ("*Bildakte*", Bredekamp 2010, p. 48ff.; the idea is traced back, among others, to Kjoerup 1978). This is to say that the nature of pictures is approached in terms of action-theoretic categories, meaning that an actor (designer, exegete, audience, parish, crowd, or any other institution) is required who

63 Translation of the quote: „Niemand bezweifelt, dass Teile von Bildern zur Gesamtbedeutung beitragen und dass sie als Teile etwas darstellen können. Was aber nicht existiert, sind allgemein akzeptierte Methoden, die erlauben, Bilder in bedeutungstragende Einheiten zu zerlegen" (Steinbrenner 2009, p. 285).

64 Translation of the following quote: „daß Bilder allein durch Vergleichsreihen […] als solche zu erkennen sind" (Bredekamp 2010, p. 29).

constitutes a picture (as a co-actor, accomplice, dialogue-partner, etc.), while the effects (appeal, authority, conflicts, etc.; see also Bredekamp & Schneider 2005) of this process are assigned to the picture as if it were the picture, which (or shall I say *who?*) brought them about as a (quasi?) actor. This equally difficult approach apparently suggests that "pictures are not acceptors, but evokers of perception-related experiences and actions"[65] (Bredekamp 2010, p. 326); "a picture speaks, and by expressing itself, it calls for a reaction"[66] (Bredekamp 2010, p. 17). On that assumption, Bredekamp compares the effects of pictures to the effects of spoken words, which, under certain conditions, count as veritable *actions* in speech act theory. However, and this is the point where I cannot follow this analogy, in speech act theory (cf. Pelz 2005, p. 248) as well as in early transfers to pictures (cf. Kjoerup 1978, p. 56) spoken words (and arguably pictures) can count as *actions* but not as actors. With this, it would remain true that actors can have variable and far-reaching effects on themselves and their surroundings (as described by Bredekamp 2010) *by virtue of* words or pictures; however, the inverse case would be excluded, namely that every source of grave effects (weather, financial crisis, etc.) is an actor-like entity. This simply collides with the intuitions on accountability, for I can hardly consult the weather, the financial crisis or a painting regarding its intentions; I would ask a painter, though. In any case, this approach, which is surely extreme in its emphasis on personification, draws heavily on the *effects* of pictures, which indicates their defining range of powers. Instead of going for properties as distinguishing marks, this account thus begins with the capacities right away, as those and perhaps only those would distinguish the picture on a wooden canvas from an array of scratches and splinters on the very same canvas. Then again, it is also clear that the notorious dichotomy between pictorial and other symbol systems does not even come into sight as a problem within this framework, since one can hardly insist that the ascribed effects are *exclusively* evoked by pictures:

> "The theory of the picture-act is therefore neither considered as a historical nor as a conceptual vote for pictures in the competition between pictures and language, but as a contribution to the reinvigoration of both in the era of challenges from the visual side"[67] (Bredekamp 2010, p. 55).

---

65 Translation of the quote: „Bilder sind nicht Dulder, sondern Erzeuger von wahrnehmungsbezogenen Erfahrungen und Handlungen" (Bredekamp 2010, p. 326).

66 Translation of the quote: „Das Bild spricht, und indem es sich äußert, fordert es [...] eine Reaktion" (Bredekamp 2010, p. 17).

67 Translation of the following quote: „Die Theorie des Bildakts versteht sich daher weder historisch noch begrifflich als bildbezogenes Votum im Wettstreit zwischen

This, of course, just shifts the problem – even more so since the 'competition' is still invoked as a motivator. Nevertheless, Bredekamp's contribution (esp. Bredekamp 2010) allows us to pinpoint two aspects with which I would like to close this section.

First of all, the survey has shown – as a whole as well as in its details – that the category 'ontological account' (which has given a title to this section) runs the risk of becoming a conceptual blank spot, since its supposed instances (the surveyed accounts) either build on an *implicit* strategy of conceptualisation, or on a *compilation* of strategies. Special credit is given to the compilation, as it seems, although one rarely finds explicit lists (cf. e.g. Kötter 2007; cf. 2.1.2.4). Instead, one finds a mixture of typified compositional traits, such as "relations of separation and overlap" (Lynch 1988, p. 225), and other criteria, pertaining, for instance, to the cognitive *genesis* (Giaquinto 2007) or social *impacts* (Bredekamp 2010) of picture-like means. The common thread, which is left in defence of the category 'ontological', is thus basically the premise that it makes sense to keep up the idea of pictures as being specific kinds worthy a distinct research agenda (cf. 4.1.3).

When I criticised the previous accounts, I did it in view of the initial problems (cf. 2.2.1.3), which require a synopsis of the functional programs *and* the functionalised means. This concerns the second finding: it turned out that, to tackle this compound, a *four*-part conceptual analysis is needed, which allows for *pictorial* and *linguistic means* (requiring conditions of application for both classifiers individually and in comparison) as well as both of their respective *function-ascribing contexts*. However, it also became clear that the surveyed accounts do not fit the entire bill, as they are inclined to just one or two of the conceptual variables. And this is in fact a strong (if not the strongest) reason for choosing Perini (2002ff.) as a reference frame, as she is the only author, as far as I can see, who tackled all variables in a manner adaptable to recent aesthetics (esp. Goodman 1976) *and* scientific visualisation (esp. to the fields of biology and systems-biology; cf. 4.2). But before coming to the detailed survey, another set of accounts will be reviewed.

4.1.3 Functional accounts: analysing practical embeddings

This last part of the survey focuses on accounts that rest strongly on the analysis of a *function* or *scope of functions*, when it comes to specifying the notion of

---

Bild und Sprache, sondern eher als Beitrag zu deren Stärkung in der Ära ihrer visuellen Herausforderung" (Bredekamp 2010, p. 55).

picture-likeness. This may be regarded as a distinct strategy of conceptualisation or as a sub-type of the ontological strategy, given that the latter is identified by the assumption that images are a distinct type of entity; this is indeed shared by the functional accounts surveyed below. Yet, unlike the foregoing accounts, which analyse the essence of picture-like means in a way that allows for their variable functional involvements, a functional account of picture-like means defines the latter as *specific* instances of *a specific* functional program (cf. Cummins 1975; cf. 2.2.1.2). The notion of picture-like means would thus belong to the same category as the notion of *medication* (which is defined against the program of *healing*) or the notion of a *vehicle* (which is defined against the program of *transport*). Strictly speaking, the type of account that I have in mind would thus also exclude what some might regard as 'classic' examples of this type of account, namely the views of H. Jonas (1961) and E. Cassirer (2009), who conceive of pictures as means that require and fortify a distant, reflective and thus human-specific relation to perceptual experiences (cf. 2.3.3; cf. Bredekamp 2010, p. 326). However, in the contexts of their work, the *genuine function* of pictures is referred to as an indicator of the human potential to establish such a relation and not as a distinguishing mark or condition of picture-likeness in the first place.

In order to substantiate this difference as well as the pros and cons of the functional account, as I define it, I will now survey a list of exemplary works. However, it would be fair to add in advance that, as with the sections above, the leading category – 'functional accounts' – offers only an approximation, because the surveyed accounts defy a strict categorisation due to their *implicit* or *combined* strategies of conception. This is particularly true for the first approach mentioned.

W. M. Ivins (1969), another pioneer of the debate, has conceived of picture-like means first and foremost as *means of communication* and, more precisely, as a distinct type of '*pictorial*' *statements* (cf. Ivins 1969, p. 1). This is also where he digs deeper for their capacities and so-justified usages, such as the exact duplication and multiplication of such kinds of statements ('repetition', as he calls it; cf. Ivins 1969, p. 1). In contrast to other approaches (cf. 4.1.2), the claimed suitability of pictorial means is also substantiated by means of *comparisons*: "In the museum I learned the bitter way how inadequate words are as tools for description, definition, and classification of objects each of which is unique" (Ivins 1969, p. 51). The corresponding merits of pictorial means are explained with regard to their properties. Unfortunately, however, this is also the point where Ivins slips into what could be referred to as a 'naïve view' (cf. 4.1.2):

> "The only set of sense awareness for which we have succeeded in making nets that catch the personality of objects are those of vision, and even they catch only a certain

> portion of it. This method of symbolization is the making of pictures or images, which unlike spoken words, are apprehended through the same sense organs, which give us the awareness we try to symbolize. Practically speaking, the visual image is the only symbol we have that does not necessarily require the translation of a sensuous awareness into terms of some other associated sense awareness or else of some extremely limited, arbitrary, and artificial convention of correspondences" (Ivins 1969, p. 53).

The arguments put forward against this view take account of the idea that the use of pictorial means is a matter of agreement (resp. convention) (cf. 4.1.2, Kemp 1990). In this context, it is also objected that, especially in scientific practices, pictures are also used when it comes to visualising entities that cannot be seen in reality, such as phenomena on the micro- and nano-scale or non-optically dimensioned features or ideas (cf. e.g. Heintz & Huber 2001, Heßler 2006). Last but not least, some would argue that the conceptual fusion underlying the phrase 'pictorial statement' is mistaken in itself, because pictorial means cannot *be* (or be used *as*) statements or propositions (cf. Steinbrenner 2009, p. 285; for a defence of this notion; cf. Topper 1996).[68] J. Steinbrenner presents a concise version of this last argument, recalling G. Frege's notion of propositions, according to which the latter must (among other conditions) entail a subject being subsumed to a predicate (Frege 1998); this, however, would not be the case in terms of pictures, which is also why they are not truth-apt, because the predicate 'true' is only applicable to propositions (cf. Steinbrenner 2009, p. 285f.). Unfortunately, Ivins has nothing to offer against this particular objection. What he offers is the assertion that pictorial means are just another *type* of statement, since they also convey information, which makes them *statement-like*. However, what makes them different is that this information is (1) addressed to a different – namely optical – receptor system (which makes them different from a "collocation of words [...] addressed to the ear", Ivins 1969, p. 61ff.), and (2) conveyed without a syntax (resp. "dictionary definitions", Ivins 1969, p. 61), which could help decode pictorial statements in the same way as verbal statements are decoded (cf. Ivins 1969, esp. p. 50ff.). While this set of criteria entails some key words employed by N. Goodman at about the same time (esp. the emphasis on syntactical differences), it leaves open how statements differ from visual experiences, especially due to the merely exclusive second criterion, which encapsulates what is *not* the case (cf. 2.1.2.3). Perhaps this is why his view has no influence beyond stimulation on the epistemological agenda.

---

68 L. Perini (2002ff.) considers this objection, which is another reason for choosing her approach as a point of departure (cf. 4.2).

His work is still worth mentioning, since “Ivins clearly establishes a correlation between changes in image-making practices and changes in science” (Perini 2002, p. 19), and thus began to examine their role and impact in times and places where they “have customarily been overlooked” (Ivins 1969, p. vii; cf. 4.1.2).

The more art-orientated discussion at this time was preoccupied, among other issues, with the digestion of Wittgenstein’s inputs, which was supposedly also the departure point of K. Walton (1973ff.). Walton draws on the much-debated notion that the meaning of an uttered phrase is defined by its use (cf. PI § 23, § 43), which he transfers to the use of all sorts of representational means, and upon which he builds a functional account of them. The novel idea underlying this account has been encapsulated by Steinbrenner (2009): “The key point of Walton’s view is that he saw the treatment of pictures – and especially the talk about them – as being their ‘essence’”[69] (Steinbrenner 2009, p. 295). Taking this as the basic idea, the task of conceptual framing opens up a set of follow-up problems, such as, *What kind of functional context gives shape and sense to these activities?* And, *What functions do they meet in respect of it?* Walton (1990) starts by clarifying that the associated context is imagination (cf. Walton 1990, p. 11ff., Steinbrenner 2009, p. 294, Frigg 2010, p. 114). In *this* context, a picture could be defined by its function as a trigger (so-called ‘prompter’) of the imagination or as an in-game requisite (so-called ‘prop’) or both. In Walton’s terms, “I propose regarding pictures as props in games of make-believe” (Walton 1973, p. 300), which implies that they may also (but need not; cf. Frigg 2010, p. 114) function as prompters, for the latter are “boon to collective imaginative activities” (Walton 1990, p. 22). Accordingly, the idea of pictorial means (and representational means in general) is not just closely tied to, but firmly defined *as* “the social function of serving as props in authorized games of make-believe” (Frigg 2010, p. 115). In terms of the mastery of these games, Walton reveals a position that seems to be compatible at least with the view of them being established (perhaps not entirely [Walton 1990, p. 301], yet still partially) on the basis of conventionalised pre-definitions:

> “In general, it seems to me that an essential part of what it is for certain objects to function as representational pictures for a group of people is that, at least ordinarily, the people know (have mastery of, have internalised) the rules of make-believe which make certain proposition make-believedly true in virtue of properties of those objects, and know them prior to acquiring the knowledge of those make-believe truths

69 Translation of the following quote: „Der Clou von Waltons Überlegungen besteht darin, in unserem Umgang – und das heißt insbesondere in unserer Rede über Bilder – die ‚Essenz‘ von Bildern zu sehen“ (Steinbrenner 2009, p. 295).

on noticing or being exposed to the relevant properties of the objects" (Walton 1973, p. 307).

He also seeks to capture the idea that a system of rules and instantiations can be *explored* (i.e. used for acquiring information) from a certain degree of complexity on, since after having established the rules of the game, changes of the instantiations or their contexts, or the sheer number of instantiations and their interrelations, can be absorbed as new inputs to the game (cf. Walton 1990, p. 13ff.). Yet, while capturing this intuition seems to benefit the compliance of the approach with recent practices, it is still necessary to account for different *types* of representational systems. According to an early specification of the "pictorials games" (Walton 1973, p. 304), "[r]epresentational *pictures* are distinguished from depictions of other kinds largely by which of their properties are responsible for what they depict" (Walton 1973, p. 314). In terms of pictures, these are, for instance, the "markings on the surface" (Walton 1973, p. 314). This, I think, benefits the strength of his approach (cf. 4.1 intro; 2.2.2.2) insofar as he specifies the *rules* of the game rather than the toys (resp. props) used in them, which are manifold enough to defy a coherent taxonomy (cf. 2.1.1). He discusses these rules more thoroughly in his later work. Yet, in fact, he does not add much to the early view, besides stressing that, "pictures are props in games that are *visual* in several significant respects" (Walton 1990, p. 293). This is certainly not a sufficient condition in times where (esp. written) language is considered as being visually dimensioned, as he admits immediately (cf. Walton 1990, p. 293). The rescue attempt draws much on the assertion that, "[l]ooking at a picture (in games of the sort in which it is its function to serve) is part of the content of the imaginings" (Walton 1990, p. 294). If this means to say though, that games of the 'pictorial' type always involve the rule that one needs to look at a picture as if one actually perceives what one makes oneself believe one sees in, as or with it,[70] this may equally count for the plastic arts and architecture. On the other hand, it will not necessarily allow for visual analogies, since I will hardly make myself believe that I am facing an intra-cellular decision when inspecting Waddington's 'epigenetic landscape' (cf. Fig. 33); a successful hermeneutic approach requires the understanding that this is precisely not the case. Yet, since visual analogies and other visual elements of modelling coin a significant part of the scientific practices under consideration, I prefer Goodman's and Perini's accounts as references (and try to defend them against Walton's notable provisos; cf. esp. 4.4).

70 For an in-depth analysis of the concepts of *seeing-in*, *seeing-as* and *seeing-with*; cf. Alloa (2010).

The last approach to be reviewed in this section has been offered by O. Scholz (2004), who takes account of both the notion of *picture-games* and the notion of *successful game-plays*. Thus, on the one hand, he welcomes the idea that something is a picture just if it is used as one. On the other hand, he makes us aware of a range of different conditions and requirements linked with the idea of success and failure in this connection (cf. Scholz 2004, p. 157). These conditions and requirements comprise nine "points or parts of accomplishment" ("*Leistungen oder Teilleistungen*", Scholz 2004, p. 166), which are all associated with the notion of having (successfully) understood a picture: The first condition amounts to (1) perceptual understanding ("*Perzeptuelles Verstehen*", Scholz 2004, p. 169). That is, in order to be able to understand a picture, it must be possible to assimilate it optically, which is in turn dependent on conditions such as the size of the picture, lighting, etc. Another basic condition is (2) the understanding of plasticity ("*Plastisches Verstehen*", Scholz 2004, p. 172), meaning that a picture is to be understood as a composed entity. Understanding a picture also requires (3) understanding the composed entity as a symbol ("*Etwas als Zeichen verstehen*", Scholz 2004, p. 172), for which it needs to show a trait (inscriptions, etc.) or shape that allows for associating it with a symbol system. Then comes what is longed for and what Scholz actually considers as a distinct condition, namely (4) understanding a symbol as belonging to a pictorial symbol system ("*Etwas als bildliches Zeichen verstehen*", Scholz 2004, p. 173). Yet, most disappointingly, this is also the point where he draws in completely (i.e. without notable modifications) on Goodman, and delegates this particular condition explicitly to his treatment. The reason why it is worth continuing with the review of Scholz's approach from this point on is simply that he reveals details about further conditions, arguably in compliance with Goodman's view, which have not yet been brought into connection with the question, *What constitutes a picture-like means judged from a hermeneutical angle* (cf. also Scholz 1993, 1998)? According to Scholz, the fifth condition would be (5) the understanding of the subject (French: *sujet*) of a pictorial symbol system ("*Verstehen des Bildinhalts*", Scholz 2004, p. 174), that is, to understand whether it is a picture of a horse or a picture of a flower, etc. Yet, understanding the subject is still different from (6) understanding the 'denotation' that is (resp. *was* or *is to be*) accomplished using the picture ("*Verstehen des denotativen Sachbezugs*", Scholz 2004, p. 177); this requires, for example, that a picture of a horse can be identified as being a portrayal of a certain horse. A distinct point is also devoted to (7) understanding non-denotative types of reference ("*Verstehen nicht-denotativer Bezüge: Exemplifikation und Ausdruck*", Scholz 2004, p. 183); as given, for instance, when one exemplifies rather than portrays the commodities sold in a store by means of the pictures in a catalogue or shop window. This is again differentiated from (8) understanding the communicative function of a picture ("*Modales Verstehen: Erfassen der kommunikativen Rolle des Bildes*", Scholz 2004, p. 187), especially that a picture shall have a certain effect on the viewer, such as being a warning sign. Last but not least, Scholz even distinguishes the condition of (9) understanding what is subtly conveyed ("*Verstehen des indirekt Mitgeteilten*", Scholz 2004, p. 188); for instance, understanding that the picture can be construed as a provocation, exaggeration, etc. Now, having completed this list of conditions, it only remains for us to appreciate that Scholz therewith "draws attention to the fact that the question, *What does it mean to understand a picture?* can be reformulated differently depending on the context"[71] (Steinbrenner 2009, p. 297). With this, he gives clear support to the notion that the use of picture-like means is much more complexly framed when put under a close-up – especially regarding required conditions and entangled functional programs, which profile and justify the so-established game. On the other hand, the impression remains that Scholz evades some of the key problems under consideration. At least, there is not much said about the exact conditions, which must be fulfilled by "a group [of entities] in order to host pictures [amongst them]"[72] (Scholz 2004, p. 188); except, of course, that it is of eminent importance to understand precisely these conditions, which comes close to a *petitio principii*. Moreover, as a pre-preparation to his list of conditions, Scholz distinguishes between the contexts of design and use (cf. Scholz 2004, p. 141ff.), which seems to imply that the design of pictures is not part of their defining use. This, however, presupposes that the rules or modes for approaching a picture have already been established, prior to the first approach ever. To this, I shall lastly object that the inspection of exemplary practices has shown that the design of an object is crucial to (if not identical with) the employment of pictorial means in the scientific contexts under consideration (esp. in the second case [cf. Falk et al. 2009] where one hardly starts or stops with a finished graphical rendition; cf. 3.3.2, 3.2).

Now, having finished the analysis, the opportunity shall be used to close this sub-section with some final remarks.

First of all, although there is no equally solid inventory of approaches in support of the category 'functional account', it would be fair to say that those surveyed

71 Translation of the quote: „Oliver Scholz' Verdienst ist es gezeigt zu haben, dass die Frage ‚Was heißt es ein Bild zu verstehen?' je nach Kontext ganz unterschiedlich lauten kann" (Steinbrenner 2009, p. 297).

72 Translation of the following quote: „[w]as […] alles dazu [gehört], dass es in einer Gruppe Bilder gibt" (Scholz 2004, p. 188).

supply a missing piece with regard to the four-part conceptual challenge revealed in the last sub-section (cf. 4.1.2). In particular, they shed light on the functional programs, where pictorial means occur, either in terms of a functional category (cf. Ivins 1969, Walton 1973, 1990), or as any functionalised entity that is part of a certain picture-prone set or setting of rules and requirements (cf. Scholz 2004, p. 158). This tackles a missing variable insofar as 'ontological accounts' (reviewed above; cf. 4.1.2) often focus on scientific game-plays, which they try to model via inspecting specific pictorial practices, without having a distinct conceptual apparatus for the identification of targeted entities. The functional accounts thus add concepts and vocabulary in support of the alignment, claimed between (what I distinguished as) picture-like and picture-based practices, since their particular strength lies in encapsulating conditions of *picture-likeness*.

In this context, however, some problems concerning the significance and strength of the accounts have been detected. One of those problems is to balance different degrees of abstraction. While 'operational accounts' (cf. 4.1.1) tend to fragment the field of concepts due to being overly specific, the 'functional accounts' run the risk of being overly general. For instance, in the case of Walton's account, the defining function is also typical of other kinds of semiotic systems. The task of specification, however, is not given priority (or even delegated; cf. Scholz 2004), although it is indispensible when it comes to tackling the problem of and why certain means are actually capable of meeting the related functional requirements. Hence, one thing is clear (esp. 2.2.1.2; 4.1.2 intro): in order to expound the function of an entity (e.g. of a medication) it is not only important to reveal the defining tasks for the accomplishment (e.g. treating an illness), but also the capacity of the medicament's components (e.g. killing a certain type of bacteria) in consideration of the capacities required to accomplish this task (e.g. killing a certain type of bacteria).

On that note, the impression is left that, considering the four-part challenge opened up by the research agenda of this thesis, the functional accounts (also) delegate a significant part of the answer to later treatment.

4.1.4 Interim summary

To conclude the literature review, I would like to give a brief account and summary of the findings drawn from the foregoing survey. The findings vindicate the choice of the reference approaches that shall be discussed and modified in the next chapters (cf. 4.2, 4.3).

However, first of all, I shall again mention that the survey is by no means comprehensive. Although I have argued against Elkins (2011) that it is possible to survey at least the *strategies* of conceptualisation (cf. 2.1.3), this does not imply that it is possible (or even necessary) to consult every instantiation of them, since there are many combined or implicit strategies in the literature (cf. 4.1.2, 4.1.3), and since it should be possible to find alternative ways of distinguishing between strategies (cf. e.g. Bisanz 2002). I would rather insist that the surveyed accounts already help to elucidate *why* some of them are more promising than others, and perhaps also *that* there are not many promising ones.

Additionally, there are definitely further strategies that have not been explicitly introduced above because they will face similar problems. For instance, I have not involved the whole tradition of defining picture-like means by reference to their *psychological effects* (cf. e.g. Gombrich 1977, Hopkins 1998, etc.), because I think that effects in general, and psychological effects in particular, are difficult to discern and trace, for they still defy a causal modelling; a call for complementary specifications is thus to be expected. Moreover, the semantic field associated with the terms 'picture' and 'image' has survived the rise and development of psychology so that a correlation to a *certain* state of this discipline will lead to anachronisms and other difficulties when applying the resulting concepts (esp. also to scientific practices, where the psyche is certainly an influential, but hardly generalisable and trusted factor; cf. Daston & Galison 2007).

As regards the *surveyed* approaches, it can be added as a general comment that they are diverse enough to unmask the use of the terms 'discussion' or 'debate' as following the idea(l) of a coherent discourse on the epistemic weight and surpluses of pictorial means. Yet, in fact, the authors rather seem to work on different parts of the conundrum, which concerns the major finding of the review: in order to tackle the initial questions, a set of four variables has to be dealt with: (1) picture-likeness, (2) language-likeness, (3) picture-basedness (esp. in science), (4) language-basedness (esp. in science), although hardly a single one of the accounts attempts a synthesis (or at least synopsis) of all of them. This is needed, since, as was best summarised by the analogy in 4.1.3, in order to expound a certain function (e.g. of a medication), it is not only important to reveal the tasks to be accomplished (e.g. treating an illness), but also the capacity that a medication has, thanks to its components, etc. (e.g. killing a certain type of bacteria). Consequently, when it comes to comparing two opposed functional programs and means, another pair of tasks and means is to be analysed, yielding the four variables as noted above and elsewhere in the survey.

Other more detailed observations pertain to the strength and soundness of the particular strategies and associated accounts, respectively. In this context, I have noted on several occasions that the employed sorting system merely builds on analytical categories. In practice, we do not only find, but also profit from *combined* strategies (e.g. from conceiving of pictures as facts and acts; cf. Bredekamp 2010, p. 48; Boehm 2001, p. 13). This became particularly evident in the case of the 'operational accounts' (represented by the rather extreme approach of Bridgman), which made clear that a *pure* focus on operations cannot meet its own standards, especially concerning the operationalisability of its own conceptual footing. By contrast, the 'ontological' and 'functional' accounts could only be blamed for not capturing the entire semantic field (which mainly concerns the notion of picture-likeness) or for not allowing for significant practices (which mainly concerns the notion of picture-basedness). Other objections, especially those pertaining to missing information on the four variables, can only be formulated against the background of the research agenda set up earlier in this thesis (cf. 2.2.1.3).

Like the review of terminological variances (cf. 2.1), the concept-focused survey thus yields a range of virtues to guide the choice of test-run views and frameworks. First of all, there are some truisms (or "semiotic trivialities" ["*semiotische Trivialitäten*"] Schirra 1999, p. 104) that come up in the debate: pictorial means are *symbol systems*, which are deeply entangled *with* and thus essentially constituted *by* their contexts of use, and which, according to the logic of *means*, are employed in more or less successful ways.

To this composite sketch, I would like to add that a qualified reference framework is expected to allow for the diverse (and perhaps still diversifying) scientific practices that require, among other points, a refraining from the unwary projection of aesthetic categories, such as of the 'finished oeuvre' or the related separation of 'design and use'. What is more, if the dichotomy between pictorial and linguistic means is taken seriously as a motivator, the framework will be expected to account for such a distinction in a way that accounts for their capacities and that also captures underlying differences (e.g. between diagrams, graphs, etc.) – since otherwise there is no aid to casuistry. Last but not least, a point not yet mentioned: there is a tendency to defend the notion of picture against *any kind* of specification (cf. e.g. Schich 2009), as if there were a natural correlation between conceptualisation and hermeneutical consolidation. Yet, since there is no evidence that research on pictorial means corresponds to a modern cosmology, seeking to capture all that there is in the world, I will just take this as another plea for a well-versed and context-adaptive account that does not neglect salient details.

4.2 An extended argument for the epistemic weight of picture-like means

The previous chapter focused on "suggestions for bringing the visual into view" (Carusi 2012, p. 108), particularly where it concerns the use of visual means in science. Continuing from this point implies that I do not consider the conditions of having epistemic weight as identical to the conditions of being technically integrated into scientific activities or reports thereof, which is true for all sorts of things and acts, including the use of white boards, coffee and other office utensils. Certainly, this is not to say that coffee, for example, has no supportive value at all. In fact, one may conceive of it as an auxiliary tool depending on one's individual research schedule and bodily preconditions. The supply of coffee is apparently just not specific enough to be put on the spot when it comes to the core of research-related concerns, routines and self-conceptions.

Why mention this? The point is that for a long time, according to the foregoing literature reviews, visual (and esp. pictorial) means used to have the status of coffee in that sense, that is, they were used to sharpen the view, channel concentration and to assimilate more inputs in a shorter time, etc. But if somebody had asked for the key ingredients of a scientist's work, one would not have started with such pragmatic details; just as one would not define the Olympic Games with reference to sportswear, although such textiles or other equipment might surely help to explain individual successes and records.

One could say, if one wanted to be generous, that the trend towards marginalising pictorial means in this manner was counteracted whenever they were methodologically specified and assigned a technical name that alluded to specific involvements (e.g. 'visualisation' or 'illustration'; cf. 3.3.3). In terms of the Olympic Games, this would be the point where the use of pictorial means would become accepted as an Olympic discipline, rather than as a contingent crutch. Integrating these means into scientifically justified methods thus already corresponds, in my eyes, to crediting them with *epistemic weight*, assuming that it implies the disputable belief that the means can play a functional role in the corresponding modes of proceeding, which in turn can be involved in scientific projects, programs, policies, etc. (cf. 2.1.2.3). Conversely, it would at least seem that the capacity of the large-scale institutions could be traced to the performances of the integrated parcels.

More importantly: from this wide-scoped definition and the literature reviews, which proffered all sorts of picture-based measures – from the constitution of research objects (cf. e.g. Lynch 1988; also Rheinberger 2001) to the proof of geometrical theorems (cf. e.g. Giaquinto 2007) – it follows that a *choice* is to be made here, in order to narrow down the analysis to a set or type of attention-worthy methods.

As announced earlier in this thesis (cf. 2.1.2.3), the choice came down to the method of *argumentation*. Thus, the focus will tentatively be narrowed down to those entities that have the capacity to become functional parts of arguments. If *this* can be said of pictorial means, they will be attributed 'epistemic weight' in the context of this thesis.

The two main reasons behind this choice are (1) that the method of argumentation has been at the centre of attention in philosophy of science, especially also throughout its language-focused period (as characterised above; cf. 2.3.2, 2.3.3). Therefore, if certain in-stances meet the condition of being *picture-like* and being an *integral functional part* of an argument without relying on a stipulation of these notions, they deserve the interest of both – those who tackle the empirical or psychological foundations and those who tackle the epistemological foundations of scientific knowledge (and those, of course, who actually agree on this dichotomy [cf. e.g. Giaquinto 2007, p. 1], which will be quoted here, but not defended). The other reason is (2) that the context of argumentation became an issue rather late in the day (cf. Giere 2007, Topper 1996, Perini 2002ff., Giaquinto 2007; cf. 4.1), so that there are more things to say from that angle.

The goal of this chapter accordingly is to devise a framework with reference to which it is possible to judge assertions about the epistemic weight of pictorial means as true or false in *this* sketched sense. To this end, the related truth conditions must be specified from an epistemological (cf. 4.2.1) as well as semiotic angle (cf. 4.2.2). This will happen with reference to the previously mentioned account of L. Perini (2002, 2005 and 2012). The limits of her account will become apparent as soon as it comes to casuistry, that is, finding a stance that works also with the examples presented above (cf. 3.1, 3.2). The identified drawbacks shall be used as a starting point for a revision of her account (cf. 4.2.5). As regards the refinement, special emphasis will be given to the concept of *ad hoc syntaxes*, which, from my point of view, is the missing conceptual key in this puzzle-solving task, since this concept allows to account for the examples at stake and surpasses other suggestions in this array (which will be argued with special regards to Seja 2009). Further applications follow in later sections (cf. 4.3).

4.2.1 The role of picture-like means

Given that the epistemic weight of an entity depends on some sort of functional relation to arguments, grammar allows for more than one option after this first step of specification: One may say that pictorial means relate to arguments

(1) … in the way that they are employed as arguments (i.e. they occur 'as' arguments), or
(2) … in the way that they are functional parts of arguments (i.e. they occur 'in' arguments), or
(3) … in the way that they contribute to particular arguments without being an integral part of them *as* arguments (i.e. they occur 'in mere support' of arguments). In this third case, the function is defined against another program (e.g. 'modelling' or 'communication'), which may be beneficial – if not essential – to the substantiation or spreading of arguments, but which is neither identical to (option [1]), nor fixedly integrated into (option [2]) the series of tasks that is to be accomplished for establishing an argument.

The question now is: which of these possible relations, which I can derive from the corpus of premises, make sense against the targeted practices and the concerns of clarification and, if any, under which conditions?

4.2.1.1 Picture-like means *as* arguments

L. Perini takes into account options number (2) and (3). Number (1), however, is not even mentioned as an option. I will first attempt to explain why this option is neglected.

I believe Perini refrains from commenting on option (1), as it would not make much sense to ask if – and, if so, how – pictorial means relate to the class of arguments, if they are, albeit not at first glance, *instances* of that class. This would reduce the controversy to an odd classificatory problem, analogous to asking: "Are these entities that are commonly classified as 'pictorial means' entities that are commonly classified as 'arguments'?" This question makes sense only when one has reason to suspect that the concept of an argument is unspecific enough to host real surprises amongst its instances; however, this collides with its status in logic. Alternatively, one might suspect that arguments are known under different names, which in turn collides with the fact that the corresponding terms 'argument' and 'picture' are not treated as synonyms. The consequences of a positive answer would thus be quite irritating. One could think of reactions such as: "I did not know before that the Tate Modern is in fact a gallery of arguments. They are difficult to identify, but I will think about counter-arguments soon, etc." Thus, not many words are necessary to reveal that option (1) comes off as a brush against the lines of common parlance. And this is always difficult to defend against the suspicion of stipulation, especially since the question is whether the relation between pictorial means and arguments can be conceptualised in a way that comes closer to (and not further away from) recent practices. Moreover, the initial question was how pictorial means *relate* to the functional program of argumentation. Taking seriously the hypothesis that pictorial means *are* arguments would imply that one declares an assumed identity-relation to be a serious problem, as if it was indeed *not* assumed that there is semantic equivalence. After all, it would come close to asking: "He is [presumably] an unmarried man of marriageable age, and he was never married before, but is he also a bachelor?"[73] (Müller 2003, p. 6). If this was the leading question of this thesis, it would turn any search for its answer into an awkward project, even more so as it collides with the widespread notion that the question about the epistemic weight of pictorial means is correctly posed as a question, that is, it truly requires a synthetic judgement in response, not just an explication of tautologies.

Considering these doubts and objections, there seems enough support to put option (1) in brackets and to continue with options (2) and (3). Perini presumably did the same, as she does not even mention option (1).

4.2.1.2 Picture-like means *in mere support* of arguments

In contrast to option (1), options (2) and (3) fit the criteria of soundness at least in the sense that they can be meaningfully called into question. Perini has recently distinguished between two hypotheses, which I would like to take as specifications of these options:

> "One way visual representations might function in scientific reasoning is to convey content that is true or false, analogous to making a claim [option (2)]. An alternative way that visual representations might function is as an object that may make statements true or false, but it is not itself true or false, analogous to a scientific model [option (3)]" (Perini 2012, p. 142).

The hypothesis that I set in correspondence to option (3) is discussed later on in her work (esp. Perini 2012, 2012a) and not least in order to show that, just because option (3) can be defended as a hypothesis, this does not mean that option (2) cannot be defended (cf. Perini 2012, p. 142). Let us therefore briefly consider what she says about option (3), in order to clear this supposed obstacle out of the way, before coming to what is in my eyes the more provocative and underexposed discussion of option (2).

---

73 Translation of the following quote (from a treatise on an ethical argument, but still worthy of use here): „Er ist [vermutlich] ein unverheirateter Mann im heiratsfähigen Alter, der noch nie verheiratet war – aber ist er auch ein Junggeselle?“ (Müller 2003, p. 6).

According to option (3), to recall the initial wording, it may be the case that pictorial means relate to arguments in the way that they contribute to particular arguments without being integral parts of them *as* arguments (i.e. they occur 'in mere support' of arguments). Leaving aside possible communicational or rhetorical modes and effects of contribution – such as evoking attention or sympathy for certain beliefs or conclusions[74] – one could think of functional support in the sense that pictorial means serve as benchmarks for their own descriptions. The descriptions in turn might be used when it comes to defending the factual assertions that are extracted from them. Altogether, the assertions, the descriptions and the pictorial items can be used to create an argument:

<table>
<tr><td></td><td></td><td></td><td>Picture x (e.g. from a radar trap) fits description y of itself.</td></tr>
<tr><td>Assertion A</td><td>(e.g. 'The speed limit is 30mph')</td><td>(w)</td><td></td></tr>
<tr><td>Assertion B</td><td>(e.g. 'Paul went down the road
y is at 40mph')</td><td>(w)</td><td>|y entails B and adequate</td></tr>
<tr><td>________</td><td></td><td></td><td></td></tr>
<tr><td>Conclusion</td><td>(e.g. 'Paul exceeded the speed limit')</td><td>(w)</td><td></td></tr>
</table>

This argument continues to be an argument even if picture x is removed from the scheme. In the case of removal, the argument even remains valid, since the two premises A and B still allow for the same conclusion. The only change after the removal of picture x would be that the argument becomes speculative, that is, it would loose its soundness, but only if (resp. as long as) there is no other option for checking whether assertion B is true or not, which could also be done by asking a witness. Picture x is also not assigned a truth-value in this scheme, which is part of Perini's notion of option (3). Accordingly, the only instances that obtain truth-values are the claims, and these are only indirectly supported, meaning that they are neither co-constituted, nor formally supplemented by picture x, as one would not assign any truth-values to a tachometer or a witness, but only to related assertions.

---

74 Creationists are famous for illustrating their claims about the simultaneous existence of humans and dinosaurs with images that show these species in co-existence (cf. e.g. http://creationists.org/dinosaurs-humans-coexisted.html [checked on 16 October, 2012]). Accordingly, the claim is set in the context of images (et v.v.), which, at a closer glance, have a rhetorical or decorative function (or both), insofar as they assemble representations of dinosaurs and men or what looks like traces of them, while, at the same time, there is no acknowledged archaeological evidence in support of this fact.

According to this scheme, it seems that picture x indeed serves as what Perini calls a "truth-maker" or "description-fitter" (Perini 2012, p. 143). Perini adopts these classifiers (mainly from Thomson-Jones 2006) in order to distinguish the way of using something in mere support of its own description(s). This would concordantly imply that picture x is ranked among those items that are consulted because of their *properties*, and not, or less likely, because of the information that is conveyed by them.

Thus construed, option (3) becomes interesting in the context of this thesis, insofar as it is further substantiated when stating, for example:

> "Fig. 25 [a] and [b] [...] corroborate the claim that there is an optically discernible difference between the [...] transduction mechanisms, namely between diffusion (left) and active motor transport (centre) [...]" (cf. close-up analysis of Fig. 25; cf. 3.2.10).

And perhaps also:

> "Fig. 12 (esp. [b] and [c]) further supports the claim that the stable solutions change with the supply of stimuli. This correlation is drawn from (and warranted by) the variant sizes of the areas covered by lines in (a), (b) and (c)" (cf. close-up analysis of Fig. 12; cf. 3.1.6).

These quotations indeed cover the two semantic fields of (1) drawing implications from the description of a figure and (2) guaranteeing that the implications could actually be drawn from the description of that figure.

On such a bases, it is then argued (e.g. by Goodwin 2009) that the use of figures corresponds to the use of *models*, given that "a scientific model is an object, about which certain claims are true or false, depending on whether or not the model fits the description conveyed by the claim" (Perini 2012, p. 143; this definition is aligned with those offered by Goodwin 2009 and Giere 2006). Accordingly, if this reading of option (3) is adequate, pictorial means belong to the class of *models* in this sense. But of course, in order to prove this to be the case, it must be demonstrated that option (3) is sound and significant enough to allow for all facets of picture-aided practices in science.

In this context, Perini argues as follows. First of all, she assumes that pictorial means are entities that *exist* – in the most basic and pre-sceptic sense of this term, that is, irrespective of their mode of existence (as possible or actual entities; cf. Richter 2011), they form a part of our world. And for this reason alone, "[i]t is trivially true that any visual representation will make some claims true; each will fit a correct description of its visible form" (Perini 2012, p. 146). What is more, *some* of these claims, she figures, are indeed used in scientific contexts:

> "For example, when scientists investigate protein structure they often use images produced by the interaction of crystal and X-rays. The crystal diffracts the rays, which interfere or reinforce one another, producing a complex pattern of spots [...]. To derive information about protein structure from the diffraction pattern, scientists measure the localization and intensities of the spots. That yields numerical representations of the form of the image, which are then subjected to intensive analysis in order to generate a model of the protein structure" (Perini 2012, p. 146; also Perini 2012a).

Perini assumes that the involved entities (esp. the diffraction pattern) are indeed used as *visible* devices in this context, and not as *visual* devices in a semiotic sense. It can thus be maintained that they (resp. their descriptions[75]) help judge assertions as being true or false (or as being adequately gained; cf. Parker 2009), without being themselves true or false or without being in any other way prone to becoming parts of arguments.

This is perhaps analogous to the restoration of a sculpture. The latter might be brought into shape in order to be examined by historians, but it is trivially not present *in* the set of (e.g. architecturally, anthropologically or geologically focused) assertions that are drawn from (or with regard to) the *descriptions* of the sculpture. The latter is just consulted as an object when it comes to evaluating its treatment with respect to technical guidelines and conventions for preparing and describing such an object.

Other examples can be taken from botanical research. Here as well, visual means have long been involved in descriptive approaches. Take, for instance, the art of illustration:

> "[In the eighteenth century,] [t]he contrast conjured up by the phrase 'drawn from nature' was not only between reality and fantasy but also between drawing from a model or, often, models (even if these were dried, flattened herbarium specimens or bloated anatomical preparations [...]) and copying another drawing" (Daston & Galison 2007, p. 99).

Accordingly, this establishes a systematic relation between scientific models and the visual 'copies' made of them. Both instances are addressed as aids to the facilitation, storage, and multiplication of *descriptive* (e.g. morphological or taxonomic)

---

75 One may object that a naïve view crawled into this construal of option B. At least some would argue that objects are not given as clusters of features; *features* are functional parts of propositions (cf. Wedekind 1998 et al., Luft & Kötter 1994). Consequently, assertions can only be matched with proposition-based *descriptions* of objects, not with reference to objects as such. However, since Perini does not spend her full attention to option (3), I will simply mention this potential controversy and leave it at that.

accounts towards the inspected specimens. And when it comes to justifying claims about them, it would indeed seem that the corpus of reconstructed descriptions – and not the described entities themselves – is used as a source and testimony of the extracted factual assertions. After all, it seems fair to say that there is timely historical support for option (3), that is, for the hypothesis that "in some circumstances scientists use images in ways that are best described as using the image as an object (or description-fitter)" (Perini 2012, p. 146).

However, and now comes the critical part, this does not imply that it is correct to *reduce* the epistemic weight of visual means to the 'truth-making' function of figures (cf. Perini 2012, p. 146); because one cannot validly infer from a sparse and notoriously open set of examples the exploitability and methodological status of visual means *in general*. This, however, has been done – for instance, by W. Goodwin (2009), who claims that pictorial means are *rather* used in ways that comply with option (3) than in ways that comply with option (2) (cf. Goodwin 2009, p. 373; Perini 2012, p. 144). In fact, such general assertions are not only problematic in the face of scientific practice at large, which is never fully accounted for by any exemplary set of instances, but also in consideration of these exemplary sets *themselves*. For instance, "there is nothing [...] that indicates that these diagrams [of structural formulae, which he uses as examples] are not functioning to make a claim [in the sense of option (2)]" (Perini 2012, p. 145). Thus, just because something is used in ways that are best captured by option (3), this does not *exclude* the possibility that the very same thing *is* or *can be* used in other ways, too (cf. Kirschenmann 2010, p. 153; this is also supported by the notion of entangled frameworks; cf. 3.3.1). Goodwin (2009) is therefore double mistaken when he claims:

> "In any individual case, one would have to decide in advance of applying Perini's argument [on option (2)] whether the conclusion of that argument was true in the relevant case. As a result, Perini's argument can play no useful role in revealing the semantic, or explanatory role, of visual representations in science" (Goodwin 2009, p. 376f.).

First of all, as we have seen before (cf. 2.2.1.3, 2.3.2, 2.3.3), recent philosophy of science tends to refrain from top-down approaches, for the reason that science 'itself' cannot be analysed as a functional program (cf. 2.2.1.3; Feyerabend 2010, p. 7). Accordingly, there is no (and perhaps never will be a) meta-methodological concept of research which would be immune to empirical exceptions, and which would allow us to state categorically that certain tools are employed in certain ways. Therefore, Perini rightly insists on her case-based approach. After all, it is Goodwin who needs to accept that his approach is thus also doomed to an inductive logic. Secondly, as regards the *ad hominem* dimension of Goodwin's objection, I claim that on the level of *capacities*, Perini truly attempts to allow for more general traits (see below); but, again, this does not imply that case-based analyses can be rendered obsolete once and for all.

To these objections – not against option (3), but against a *reductionism* in favour of this option – I would finally like to add that, even if the conceptual proximity between visual means and models "has the advantage of bringing scientific images within the purview of a high-level account of scientific knowledge [insofar as the latter is focused on models]" (Perini 2012, p. 143), this is of no further benefit to recent discussions in the philosophy of science; in fact, the function of using something as a benchmark for its own apperception was not even doubted (perhaps only abstracted from) by the logical empiricists (cf. 2.3.2, 2.3.3). Thus, if Goodwin was right in favouring option (3), then the whole debate on visual means could be widely or fully synchronised with the recent debate on models.

4.2.1.3 Picture-like means *in* arguments

The cautious use of such terms as "rather" and "widely" indicates that the strongest objection to the preference of option (3) would be the successful defence of an *exception* to it – in this case of option (2). This leads us now to the core argument.

According to option (2), to recall the initial wording, pictorial means might relate to arguments by virtue of being *functional parts* of them. In order for this to hold true, it must be shown – according to the framework that has been established around the notion of *function* (cf. 2.2.1.2) – that pictorial means have the capacity to fulfil a task that is specifically needed to establish an argument. It would neither suffice to show that they are *meant* to be used in any such a way, since this would not prepare for getting a stance on whether it makes sense to favour them over other means (in scientific practice) or other issues (in the philosophy of science); nor would it help to expound their function against any pre- or post-preparatory functional program (e.g. writing down arguments), because then there would be no limit to the listing of functional ingredients; one could also take into account "words, subsentential expressions and even punctuation marks" (Goodwin 2009, p. 376).

In order to avoid these traps, Perini (2005) suggests approaching option (2) in terms of the problem, whether pictorial means have the capacity to convey claims – and thus either premises, or conclusions – in the context of an argument. In accordance with this modified conceptual framing, the scheme of an argument could look like this:

Assertion A	(e.g. The road traffic regulation says ['The speedlimit is 30mph'])	(w)
Assertion B	(e.g. [Picture x] shows that ['Paul went down theroad at 40 mph'])	(?)
Conclusion	(e.g. 'Paul exceeded the speed limit')	(?)

Now, what is needed for a picture x to take the assigned part in the scheme? Is picture x able to meet the corresponding conditions? One thing is clear: the answer must be carefully thought out, as the expectations multiply quickly. If something can be "a full-fledged component [...] of scientific arguments" (Goodwin 2009, p. 372), it may also be part of a confirmation (Perini 2005a) or an explanation (Perini 2005b), and perhaps also of other types of 'reasoning'. Any framework established in support of this notion would thus need to allow for weighty consequences.

In order to forestall some instant objections in this context: of course, one should not make the same mistake as with option (1); the ontological or semantic *equation* of pictorial means and premises or conclusions would surely conflict with common habits of classification (cf. 4.2.1.1), and it seems hopeless to escape these troubles, since one would need to justify why the assertion 'Pictorial means occur *as* arguments' is trivially faulty, whereas the claim 'Pictorial means occur *as* premises or conclusions' is at least debatable. On the other hand, Perini never claims anything of the sort. She seeks to clarify whether pictorial means *can convey* claims, and not whether they should be identified or treated *as* claims. This does not even appear adequate in the case of the linguistic counterparts, because it is widely acknowledged that a sentence *can* be used to convey a claim under *additional conditions* (cf. III intro), but it *is* neither yet a claim nor is it *made of* claims. Therefore, any criticism of that kind is beside the point.

It remains now for us to ask, using the words of a fierce critic of option (2): "[W]hat makes a symbol system linguistic [or pictorial], or [is there] anything essential about the role of linguistic [or pictorial] systems in the expression of claims" (Goodwin 2009, p. 374). Before coming to the first part of this question (which is tackled later here; cf. 4.2.2), let us analyse the suggested conditions for being 'essential' or 'integral' to arguments in order to approach Goodwin's question from the chosen conceptual introspective.

According to my reading of Perini's approach, which in turn builds on her reading of A. Tarski (1956) and N. Goodman (1976), Perini evades the notion of *predication*, that is, the idea of subsuming a subject under a predicate, when it comes to analysing the conditions of conveying a claim. Thus, while many would argue that a claim implies exactly this, that is, the coupling of a predicate and a subject in such a way that something is stated about the subject (cf. Frege 1998; Steinbrenner 2009, p. 285f.), Perini employs another – equally approved, but rarely added – condition. She writes:

> "[D]efining arguments in terms of sets of statements is [...] too narrow [...]. An alternative is to use a very broad characterization of an argument, as a set of representations, each of which must be capable of supporting a conclusion [...]. This characterization does not settle the question of whether or not a figure can be part of an argument, but it does point the way to an important criterion. The support that premises provide a conclusion is analyzed in terms of validity or strength, and soundness, so any representation that is an integral part of an argument must be one to which those features could be relevant. Validity, strength, and soundness are understood in terms of the truth conditions of premises and conclusions, so representations that contribute to arguments must have the capacity to bear truth" (Perini 2005, p. 263; underline added).

Some have criticised her for stressing a point of minor relevance here (cf. e.g. Goodwin 2009). However, while there are indeed critical points about the involvement of *truth*, the trouble comes not with relevance – at least not insofar as it concerns the notion of a claim, because it is a truism worthy of encyclopaedic entries (cf. e.g. Gabriel & Zantwijjk 2001, p. 436–444) that claims (resp. 'propositions' or 'statements') can themselves be treated as logical subjects, that is, that they can be subsumed under *meta*-predicates, such as 'true' or 'false'. Other entities may be judged as being *correct* or *adequate*, but they are not *specifically true* or *false*. It is to the credit of Goodwin, though, that, in order to extract this essence, some irritating notes need to be factored out, such as the call for validity, strength and soundness, which is not further specified and never mentioned again in this or any other of Perini's papers. These and other ornamentations will therefore be put into brackets.

The question is, how does this account open a door for pictorial means at all? One could expect that, after the rejection of the correspondence theory of truth, any allusion to the notion of truth is doomed to a metaphysical overtone.

In fact, the trick is simple and, as I claim, beyond reproach, although it is hidden behind a verbose excursion into early 20th century's logic. Perini (2005) focuses on the notion of truth in *formal languages*. In other words, something (preferably a claim) is true, just if it is an instance of a fixed and limited set of rules of combination (syntax) and representation (semantics). Transferred to the account of Walton (1973, 1990), for example, one could argue that a claim is true, if it complies with the program of making a claim, as set up in the associated game. On the other hand, something is *capable of conveying* a true claim (resp. of "bear[ing] truth"[76], Perini 2005, p. 263), if it is *part of* a set of instances put forward as a claim, and if it *allows for judging* whether the suggested claim complies with the rules of claiming or not. In that case, she argues, the entity is *capable* of becoming part of a claim and thus also of becoming part of a premise or conclusion.

I think that this already captures the essence of the epistemological part of Perini's framework, but let us have a look at how she develops these points in detail, in order to flesh out the associated conditions and related arguments.

The leading paradigm is "that sentences are true when the states of affairs they refer to obtain" (Perini 2005, p. 275). However, Tarski (1956) has shown that, since "the truth of any sentence depends on the fact expressed by that particular sentence" (Perini 2005, p. 275), and since there are no rules in non-formal, so-called 'object'-related languages, which would clearly and completely determine the set of reference facts, it is not possible to establish a substantial theory of truth that helps determine the truth-value of *any* sentence in non-formal, object-related symbol systems. To put it bluntly, there is no general guide to the relation between all-day language and the supposed non- or extra-linguistic facts to which language is referred to. Consequently, we will never be able to *infer* claims that are empirically true and to those that are not.

Following on from this, the idea adopted from Tarski (1956) is that in the case of 'formal' systems – that is, systems that entail a fixed and limited set of symbols and rules of combination (syntax) and representation (semantics) – it is possible to define general truth conditions and to judge recursively (i.e. with reference to these truth conditions), whether an assumed instance of that system fits the condition or not; if it fits the condition, it is *formally true*. Accordingly, in order for a sentence (or any symbolic compound) to be formally true, it must be an instance of the assigned reference system. As Perini puts it:

> "A materially adequate definition of truth allows, for every sentence s in the language, derivation of a statement of this form: X is true if, and only if, p. Such a definition is a set of statements of the conditions under which sentences of the language are true. X is the name of the sentence, and p is a translation of sentence s into the metalanguage of the definition (or p is s itself, if the metalanguage includes s).

76 Much of the current criticism is due to the infelicity of this expression. Goodwin (2009), for example, objects that not each component of a claim must be true. He is right, or course. But, as far as I understand Perini (2005), 'bearing truth' does not correspond to 'being true'. If something is capable to 'bear truth', this is just to say that it allows for being considered as formally true or part of a formally true compound.

> The name is determined by the form of the sentence: it is a function of the sequence of symbols constituting the sentence" (Perini 2005, p. 275).

In other words, if a sentence s is an instance of as assigned 'meta-language' – that is, if s can be expressed in this language, where it is thus known under the name of X – then it is formally true. And, to recall Perini's addition, it is thus also capable of becoming part of an argument in a formal or semi-formal context (cf. 4.2.5.4, 4.3.2.3).

From the viewpoint of this thesis, the question now is, what is needed for a symbol system to be capable of bearing truth in this particular sense? In this context, Perini rightly notes that "[i]n principle, any perceptible thing can be assigned a meaning. And because the content of a sentence (or other acknowledged truth bearer) can be assigned to any object, any object can bear truth" (Perini 2005, p. 274). This is to say that I can use little stones, colour dots, gestures and other sorts of entities to make up a formal game; the choice rather depends on whether these instances can easily be handled and on whether they are suited to the creation of more complex expressions. In terms of more complex symbolic *systems*, one may conclude that any of the integrated instances must allow for "showing that the state of affairs the symbol refers to is not arbitrarily assigned, but is defined at the level of the symbol system [semantic condition]" (Perini 2005, p. 276). In addition, it must "allow for the recursive definition of complex characters from atomic parts [syntactic condition]" (Perini 2005, p. 276; see below).

4.2.2 Picture-likeness revisited

I have already noted earlier that characterising the occurrences of so-called 'figures' is a complex task to accomplish. This seems to be due to the entanglement of compositional, representational and methodological levels of establishing figures *as* semiotic means (cf. 3.3.3). On each level, further variations have been discovered (cf. 3.3.3f.). Therefore, in order to be able to judge assertions about the epistemic weight of picture-like means as being true or false, it does not suffice to clarify the conditions of *having epistemic weight* (cf. 4.2.1); additionally, one needs to dig deeper into the conditions of *being picture-like*, because those two sets of conditions are exactly suspected to collide with each other, while each of them presents a veritable analytical challenge. As Perini puts it: "We cannot achieve a[n] [...] understanding of scientific reasoning until we understand the tools by which scientists present, defend, and evaluate ideas; given the prevalence of visual representations in scientific communication" (Perini 2012, p. 132).

4.2.2.1 Learning from misconceptions

On the way to defining this second, semiotic part of the framework, not every intuition proves reliable, although the previous chapters made positive use of intuitions so far. Some of them lead astray or, even worse, prevent problems from being seen and tackled at all. Perini (2010) presents a potpourri of such traps of conceptualisation, which she and others reject for good reasons, considering the following examples:

(1) "Visual representations represent visible (or at least spatial) things" (p. 133).

This belief has a rich tradition. It is subject to the work of Plato (cf. *Politeia* 596e–598e), Lessing (*Laokoon oder* über *die Grenzen der Malerei und Poesie* [1964]) and those who contributed to the reappraisal of their works (resp. of the problems they deal with). And in fact, there is no controversy about the claim that visual representations are *suited* (from a mere technical point of view) to the encoding of features or relations that are otherwise expressed using geometrical, topological or graphical descriptors. The critical point rather is that these descriptors are *not only* applied to entities that literally occur in the field of vision, but also to entities that are either not available to the eye (e.g. atoms), or that have no spatial extension at all (e.g. degrees of danger or uncertainty). Therefore, intuition (1) (about the visibility of referent entities) may state a possible surplus, but not a *condition* of picture-likeness. Other intuitions are prone to a similar kind of objection.

(2) "An ideal visual representation *has* features matching those of the referent" (p. 133)
(3) "An ideal visual representation represents *all* features of its referent" (p. 133)

These beliefs "gain plausibility from our experience of familiar types of pictures, like paintings rendered in naturalistic styles, and photographs. [...] [And] [i]f resemblance determines content, it determines accuracy conditions" (Perini 2010, p. 133). Still, if there are cases in which resemblance does not, or not straightforwardly, determine the content (see criticism on intuition [1]), and if it is also true that those cases are not automatically dismissed for quality reasons, then intuitions [2] and [3] do not amount to a "definitive feature[...] of visual representation" (Perini 2012, p. 133). One may even go for a stronger wording: there will always be options for finding similarities between symbols, referents and non-referents, starting from the fact that all *exist*. Similarities are thus neither *necessarily* relevant, nor necessarily *all* relevant to *being* a symbol – in fact, this is not even needed in the case of naturalistic representations, since even those rely on models of nature that guide the abstraction or neglect of details (as has been revealed by Goodman 1976; and also Kemp 1990). Still, those who contributed to the spreading of intuition (2) and (3) are at least right in insisting that there must be a guideline for the use of a symbol. On this detour, one may, for example, end up with the *pragmatic decision* to take four lines as symbols of four trees. In such a case, one may say that the symbols and the referents share the blurry appeal of *straightness* and *thinness* combined, and perhaps also the feature of *fourness*, given that this is correctly conceived of as a feature. However, it is not possible to infer validly from such exemplary *pragmatic choices* to the rules for determining the features that obtain a representational function or that constitute visual means. This is further substantiated when considering the possibilities to *allude* to features (as in the case of an only *sketchy* square; Giaquinto 2007, p. 14), or to *fake* the impression of something having a certain feature (as in case of *trompes l'oeils*), or to *test-run* or *revise* features (as in the case of *interactive* rendition; cf. 3.2.12).

Consequently, there is not much hope for the intuition on 'shared features' – neither as part of a general ontological, nor as part of a methodological account of picture-likeness.

The problem with this intuition, as I see it, is that it paradoxically builds upon the most momentary and variable level of approaching or establishing visual means – namely the *methodological* level (resp. the level of picture-*based* measures, as I called it earlier; cf. 3.3.1) – as if *this*, of all others, were a fix(able) point of orientation. In fact, this level depends on the state of art in all conceivable respects, for instance, in regard to the temporary agendas, gusto and technical standards in the scientific or the art games. The fate – that is the failure – of any such account may thus be anticipated, using the example of the *theory of similarity* or *resemblance* (as mentioned above), according to which a picture is by definition similar to its referent. Even Goodman, who was among the first to reject this theory (cf. Goodman 1976, esp. p. 3–6), never *doubted* that resemblance *can be* an issue in the process of quality assessment,[77] but it *need not be*, and neither is

77 Goodman (1976) takes account of the topos of resemblance, but uses each opportunity to stress that it does not amount to a necessary condition, but rather to a matter of convention or pre-definition: "Realistic representation, in brief, depends not upon imitation or illusion or information but upon inculcation. Almost any picture may represent almost anything; that is, given picture and object there is usually a system of representation, a plan of correlation, under which the picture represents the object. How correct the picture is under that system depends upon how accurate is the information about the object that is obtained by reading the picture according to that system. But

there any universal method for identifying particularly *relevant* similarities (see criticism on intuitions [1] – [3]). And if 'resemblance' just denotes the fact that two entities can be subsumed to the same predicate, this definitely holds for all sorts of entities, and also for many more than the relevant aspects of the visual means (see also above). Thus, to conclude the argument: a couple of hints have surfaced, indicating that it is not helpful to build upon conditions that are prone to variation and contextual adjustment, such as those on the methodological level.

Judging from his recent success as a role model, it seems that K. Walton (1973, 1990) made a better choice in this context, as he went for the *representational* instead of the methodological level. The problem with his account is rather that, from a bird's eye view, as established by the all-encompassing notion of games of make-believe, sophisticated bridging principles are needed in order to account for bottom-level distinctions (such as between graphs, diagrams, etc.); even the differences between picture-like and language-like games, and between hybrids thereof, escape a strong grip from the introspective of a general theory of pretence, despite its explanatory merits (cf. Frigg 2010; cf. 4.1.3).

In search of an account that is capable of illuminating the state and status of pictorial practices in science without reducing them to rudimentary schemes and patterns, Perini rather suggests using Goodman (1976) as a point of reference (cf. 4.1.1).[78] Goodman devoted his attention to *compositional* traits in the first place.

4.2.2.2 From referential systems towards symbol systems

The focus on symbol *systems* is essential. Perini prefers it to the more classical focus on the relation between individual symbols and their referents:

---

how literal or realistic the picture is depends upon how standard the system is. If representation is a matter of choice and correctness a matter of information, realism is a matter of habit" (Goodman 1976, p. 3).

78 Since Goodman's account is quite popular as a point of reference (either in competition to or as a supplement of K. Walton's approach), I will restrict the analysis to Perini's way of interpreting his work. I will mention the original version only if necessary. In other words: due to the systematic focus of this thesis, I skip the historical task of commenting on exegetical problems and controversies, because the framework must first and foremost help to find an answer to the questions about the epistemic weight and surpluses of picture-like means. Consequently, the framework need not (at least not in the first place) launch a contribution to corresponding reading projects.

> "Stopping semiotic analysis at the level of the symbol-referent pairs leaves out critical information for understanding representation" (cf. Perini 2010, p. 137).
>
> "The need to focus on symbol systems, rather than individual symbol-referent pairs, is an important and underutilized insight" (Perini 2010, p. 139).

The benefits of this focal shift, which is inspired by Goodman (1976), can hardly be overestimated. One of its greatest strengths is that it brings the rule-based framings and underpinnings of the establishment and use of symbols into the centre of attention (cf. 4.2.2.3; 4.2.2.4). Consequently, it focuses on the relation between the instances of symbol systems, and also between different types of symbol systems, without getting lost in metaphysics of reference that would rest on claims about *extra*-symbolic or -systematic facts. Other traps of conceptualisation are also avoided, such as the trap of 'ontologisation' (i.e. taking pictures as natural kinds instead of means) or 'hypostatisation' (i.e. taking the relations between symbols and their referents as manifest traits [e.g. of resemblance]). Instead, Goodman paves the way for a pragmatic view, that is, he indicates, or at least allows for the conclusion that the conditions of language- or picture-likeness correlate with specific modes or schemes of *reading* marks or inscriptions. Given that the application of *rules of en- or decoding* thus moves into the centre of interest when it comes to defining semiotic types (e.g. pictorial means), Goodman's framework has the unbeatable potential of accounting for historically variant institutions and conventions. The latter *can* but need not be invoked when it comes to guiding or justifying the abstraction of compositional traits and the assignment of meaning. Accordingly, the ultimate *decision* on the applied scheme of en- or decoding is left to context-variant (e.g. scientific) rationales and ad hoc trials. Since the latter can hardly be preconceived (which has been revealed in the previous section; cf. 4.2.2.1; also 4.2.5.2ff.), it is save to say that Goodman implements conceptual *variables* at the right (namely methodological) points of his framework.

4.2.2.3 What is a symbol system?

After this early appraisal of the chosen reference framework, I will now scrutinise the reference accounts in further detail. The questions are, what is a *symbol system* according to Goodman; how does Perini make use of Goodman's conception, and to what purpose? The following quotes should prepare for a first impression:

> "Any symbol scheme consists of characters, usually with modes of combining them to form others. Characters are certain classes of utterances or inscriptions or marks" (Goodman 1976, p. 131).

Perini adopts this set of definitions without significant modifications:

> "Each symbol system is comprised of a set of characters (each character is an equivalence class of markings – utterance or visible mark), a set of referents, and relations correlating characters and referents" (Perini 2010, p. 139).

This definition has at least two advantages: first of all, it can be further specified. One may thus distinguish between different *types* of symbol systems, for example with respect to the *types of entities* that are used as visible marks, or with respect to the rules or modes of combining or interpreting them. In other words, the definition offers conceptual *variables* that help account for the (sub-)distinction between language- and picture-like systems (cf. 4.2.2.4). Besides, it fits Perini's claim that *any* kind of entity can be used to establish a class or instance of a symbol system (cf. Perini 2005, p. 274; cf. 4.2.1.3); this depends indeed on the purpose (e.g. reaching a certain community, or distributing certain contents). Therefore, a definition and associated taxonomy of symbol systems performs best when it focuses on the *rules* of establishing such a system, and not on its concrete components in the first place.

The reward of the restriction to rules of establishing or approaching symbol systems is that any such account (here esp. Goodman 1976 and Perini 2005) preserves the option of approaching visible inscriptions differently. That is to say, for example, that one may approach hieroglyphs in terms of a text, or in terms of a set of pictures; or one may approach uttered vocals in terms of a song or in terms of a poem. An illiterate person may attempt to discover the meaning of the differently shaped spaces in-between what others classify as letters. Also, "we could use photos of different animals [...] to refer to months of the year, for example" (Perini 2005, p. 268). In the end, all this is *possible*, though perhaps not common or conventional. Nevertheless, Goodman's framework is capable of accounting for such possibilities in case they are, have been or will become conventional one day.

4.2.2.4 What makes a symbol system picture-like?

Perini's argument on the epistemic weight of picture-like means relies strongly on the possibility of distinguishing between different types of symbol systems. This possibility has been rescued along the lines of Goodman's notion of symbol systems (cf. 4.2.2.3). Moreover, she builds upon the conditions of *language-likeness* and *picture-likeness*, as defined by Goodman. To minimise the exegetical provocations in this context, I will now present only those conditions that carry her argument, even though Perini aspires to adopt more or less the entire taxonomy of Goodman (cf. Perini 2005, p. 267–273).

One of the binary oppositions proposed in support of the distinction between picture-like and language-like systems is the one between *serial* and *spatial* orderings of marks (resp. utterances or inscriptions):

> "The fundamental formal difference between visual representations and text is the sequential format of linguistic representations versus the two-dimensional spatial format of visual representations. Written words are two-dimensional characters; we identify letters by their shapes, and both height and width are relevant. However, the meaning of linguistic statements is determined just by the sequence of letters, punctuation, and spaces. […] The shapes of letters are unrelated to the referents of the words and sentences they comprise, and the spatial position of words does not contribute to the meaning of text, except in terms of the sequence of linguistic characters. And even though linguistic representations must be read in a certain direction, e.g., left to right, the left-right spatial relation is not itself meaningful" (Perini 2005, p. 263f.).

Accordingly, it is argued that language-like and picture-like systems differ in the way that picture-like systems make use of 'spatial' (i.e. geometrical, topological or graphical) variables, whereas language-like systems do not, or only insofar as it concerns 'serial' orderings. However, even though this is often found presented as *the* difference between these two types of symbol systems, it does not sufficiently corroborate the distinction between them; because, what appears to be a neat opposition could also be analysed in terms of a relation between a type ('spatial') and a sub-type ('serial'). This would imply that these conditions can be combined and, thus, do not necessarily exclude each other. This is indeed true when considering hybrids, such as comic language or hypertext codes, where the location of letters and their graphical rendition might convey meaning or contents. Still, this first opposition is already useful, since it helps discriminate between *some* linguistic and pictorial systems, and it is also capable of indicating borderline cases.

Before I introduce the other conditions that render the distinction between picture- and language-likeness more defined, I suggest continuing with a revised terminology, since I don't think that it makes sense to contrast the attributes 'serial' and 'spatial'. Indeed, symbols are lined up serially in linguistic symbol systems. Yet the point is not that pictorial systems arrange symbols differently. The point rather is that they use 'geometrical', 'topological' or 'graphical' types of arrangements, including series, *as* symbols. This is usually fused under the category 'spatial' (certainly also in Perini's account); however, it makes sense to list these ingredients individually in order to avoid confusion. Talking explicitly about 'geometrical', 'topological' and 'graphical' instances also comes closer to the description-language used in science. Besides, this does not conflict with the ideas associated with the attribute 'spatial' but restricts them to a more specific scope; otherwise one might think that *spatial* information is also conveyed by acoustic or tangible systems.

With this clarified, let us now analyse the other conditions against which the distinction between 'serial' and 'spatial' modes of en- or decoding ultimately turns out to be merely supplementary.

Goodman introduces further binary oppositions that allow for both defining and discriminating language-like and picture-like systems. These oppositions are established on the compositional and representational level, which Goodman identifies as the levels of *syntactic* and *semantic* features, respectively:

> "Goodman focused on two kinds of symbol system properties: syntactic features, which have to do with aspects of the character set of a system, and semantic features, which describe relations among the referents of a system" (Perini 2010, p. 140).

The choice of these framework components is best explicated earlier in his work:

> "Comprehension of representations requires identifying and interpreting the marks that compose them. The marks in any visible representation (text or figure) must be identified as instances of characters, and the characters determine what the marks denote. […] For example, comprehension of written English requires (1) identifying words or phrases through identifying the sequence of individual letters that compose them (e.g., identify the first mark in 'cat' as an instance of the letter c, etc.), and (2) associating the correct referent objects with particular words or phrases (e.g., between cats and the character instantiated by 'cat'). Goodman identifies several variants in each of these two aspects of the interpretation of symbols, and uses these variations to categorize different symbol systems. The syntactic criteria relate to the first part of the interpretive process: how marks are individuated and identified as characters. The semantic criteria have to do with the second part of the interpretive process: how the referents of characters are defined and differentiated" (Perini 2005, p. 270).

Whether it really makes sense to introduce a *chronology* of the 'interpretive process' (first syntax, then semantics) is not an issue here, because this questionable aspect is irrelevant to the argument. It is, however, important to note that, thanks to the distinction between the syntactic and the semantic level, Goodman creates access to a whole range of different conditions of establishing or approaching symbol systems that are mutually exclusive and that can thus be used to define a taxonomy of (sub-)systems.

The conditions that are defined in opposition to each other and that are most relevant in the context of Perini's argument are those between (1) *syntactically discrete* (digital) *vs. syntactically dense* (analogue) systems, and between (2) *syntactically articulate vs. syntactically inarticulate* systems. On the semantic level Perini distinguishes between (3) *semantically articulate vs. semantically inarticulate* systems, and (4) *semantically disjoint vs. semantically non-disjoint* systems.[79] The first opposition (1) is the most famous one: it is celebrated as the key part of a master argument ("*Meisterargument*") in picture theory (cf. Scholz 1999). The other oppositions (2), (3) and (4) might be known from quality assessment or from the definition of formal systems. The oppositions (resp. the readings of them suggested here) can now be further characterised as follows.

A *syntactically dense* ('analogue') system is organised in such a way that "[f] or any two marks [resp. utterances or inscriptions] there is another one ordered between them" (Perini 2005, p. 271; cf. also Fischer 1997, p. 90). By contrast, a *syntactically discrete* ('digital') system is organised in such a way that, for any two marks, there is *no* other ordered between them. Against the background of a pragmatic view[80] (cf. 4.2.2.2), one could rephrase this by saying that a system is *read as being dense*, if any parcel within the considered area is taken as an instance of the system, while a system that is *read as being discrete* allows for the distinction between foreground and background, while the background is not an instance of the system (although it might contribute to establishing an optically beneficial contrast; cf. 2.2.2.1). For example, in A. Dürer's *Great Piece of Turf*, every geometrically, topologically or graphically describable part within the framed area is taken as being a representational part. Accordingly, even the white (possibly blank) space in the background is assigned a referent; it represents *the* background of the meadow (e.g. the sky). If somebody were to squeeze this piece of work, or shot a bullet through it, the system gets corrupted. However, if the same happens to a map of an underground railway, the symbol system may still work, as the dis-

79 As mentioned earlier, this list of categories is not complete, neither with regard to Perini (2005), nor with regard to Goodman (1976). But the argument already works with the categories introduced. Besides, some of the skipped ones are actually problematic, which will be shown later on (e.g. 'ambiguity'; cf. 4.3.2.1).

80 Many authors (cf. e.g. Fischer 1997) argue that *computer-based* representations are genuinely discrete in this syntactic sense, for the reason that every mark can be traced to *one* signal and thus never 'fuses' with others. I will reject this hypostatisation of coding systems for pragmatic reasons (see below).

crete *parts* that contain the information (e.g. that station A comes before station B) are not necessarily affected.

Opposition (2) is also established on the syntactic level, but not identically to opposition (1). A system is *syntactically articulate* if "any mark that does not belong to two characters can be determined not to be an instance of either one or the other" (Perini 2005, p. 271; cf. Fischer 1997, p. 90). That is to say that in such a system it is possible to discern all of the employed marks and to classify them as instances of syntactic characters. This sounds somewhat similar to the condition of discreteness, but it is indeed a different feature. Take, for instance, the geometrical rendition of the cell model (cf. Fig. 22). The latter could be taken as syntactically discrete, insofar as all syntactic units are individuated as well as loosely arranged against an undefined background. One may thus discriminate each of them when traversing the model. However, since the latter multiply and intersect when zooming out of the cell, it becomes difficult to keep them apart. In effect, the system could be seen as *syntactically discrete*, but *syntactically inarticulate*.

Opposition (3) adds a pair of semantic criteria. A system is *semantically articulate* "if it is possible to discriminate between referents and non-referents of a character" (Perini 2005, p. 272). By contrast, a system is *semantically inarticulate* if it is not possible to determine whether something is a referent of a character or not. Yet, even if a system is semantically articulate, the compositional level might have an impact on whether we can assimilate the system as articulate or not – that is, whether we can find the pairs between symbols and referents (cf. Fischer 1997, p. 91). Take, for instance, a clock with a non-structured clock-face (cf. Fischer 1997, p. 91). If there are no digits on the clock-face, one can only approximate the time, but it is not possible to determine exactly to which digit the clock hands point, although it is excluded from the semantic point of view that it points to two digits at the same time.

Opposition (4) finally adds another distinction on the semantic level. A symbol system is *semantically disjoint* "if no two characters denote the same thing" (Perini 2005, p. 271). On the other hand, a system is *semantically non-disjoint* if there is a set of characters that entails variably employable operators (such as letters). "In semantically disjoint systems the identity of the object is sufficient to specify which term refers to it (if any) because there is at most one such term. [For instance,][t]he numerals and natural numbers they refer to are a disjoint system. Discursive systems [...] are not, because there are many objects which are referred to by more than one word or term" (Perini 2005, p. 271f.).

When it comes to fleshing out the dichotomy between picture-like and language-like systems, the mentioned oppositions are used as follows:[81]

> "Systems that are syntactically articulate and discrete can be contrasted with systems which are syntactically inarticulate and dense. I will refer to such systems as pictorial systems, since this category includes the kinds of images we think of as pictures (photographs, perspective drawings, courtroom sketches, etc.)" (Perini 2005, p. 273).

Accordingly, the category 'picture-like' is identified with the aspects of syntactic *density* and *inarticulateness*. Whether this can be extended to the semantic level is controversial, as it might still be possible to define distinct pairs between symbols and referents and arrange them so densely that they can hardly be discriminated or if so then in the course of test-run interplays with other instances (cf. 2.2.2; 4.2.5.2).

In contrast, the category '(written) language-likeness' comprises the criteria of being *syntactically discrete and articulate*. Yet, in order to additionally obtain a set of well-differentiated, distinctly coupled *semantic* referents, further measures are needed (cf. 4.2.5.4; 4.2.3.2). Those measures are found employed in 'formal' systems, where symbols and referents are unambiguously coupled in accordance with distinct rules (e.g. in predicate logic; cf. Perini 2005, p. 271). However, these measures are too special to be considered as compelling features of language-likeness (cf. also 4.2.5.1).

Up to this point, this has been a pen and paper exercise, a pure coupling and matching of terms, conceptual ingredients and conditions of use. The controversial part comes with casuistry and with the problems generated when adding further conditions and requirements (such as of 'having epistemic weight' or 'surpluses').

First of all, it is important to note that when taking the listed criteria and the distinction between serial and spatial arrangements together, this enhances the options of classification, as demonstrated by the following example (cf. Tab. 1):

---

81 It needs to be added here that some introduce another specification. Fischer (1997), for example, does not locate the opposition between linguistic and pictorial systems, but between *written* linguistic systems and pictorial systems. I think this is a sensible move, since what is true for a textual inscription (e.g. in terms of discreteness) does not automatically count for the oral rendition of the textual inscription – one of them can be articulate, the other not. But since Perini also uses merely printed examples (and I do too), I think we can adopt this specification without facing a change of the argument.

Tab. 1:

Much ado about nothing	*x (Px)*	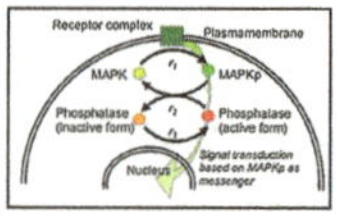	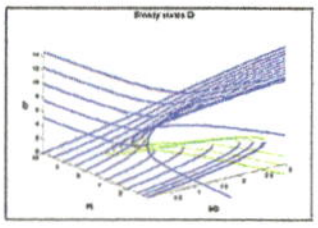	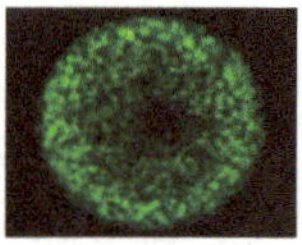
		Fig. 24	Fig. 13	Fig. 25
Serially arranged, syntactically discrete and articulate, semantically non-articulate	Serially arranged, syntactically discrete and articulate, semantically articulate	Spatially arranged, syntactically discrete and articulate, semantically articulate	Spatially arranged, partially dense and inarticulate, semantically articulate	Spatially arranged, syntactically dense and inarticulate, semantically non-articulate
language-like	**language-like**	**(rather) language-like**	**(rather) picture-like**	**picture-like**

There might be surprises amongst these classifications. For example, what seems to be *picture-like* in respect of the dichotomy between *serial* and *spatial* orderings – namely the diagram (Fig. 24) – turns out rather *language-like*, when taking into account the condition of syntactic discreteness. Perhaps equally surprising, yet tending towards the opposite direction, the graph (Fig. 13) is classified as 'rather picture-like'. This is because the blue and green curves are densely composed; yet, still, the system builds on a distinct semantics, as each atomic part represents exactly one value (cf. Perini 2005, p. 271; 4.2.5.1). The microscopy-like image (cf. Fig. 25), in turn, is classified as a genuinely picture-like means. In this connection, it should be important to add that the classification works without, and *only* without, information about the computer-based or -generated environment of the figures. In fact, I firmly insist that the technical origin *should not* make a difference in the classification (which has been argued, e.g., by Fischer 1997). Whether or not something is construed *as* being syntactically dense depends solely on the agenda pursued (resp. on the coding scheme employed). Certainly, it might be that a pixel is considered as a syntactic class, but this is not true for every case. The cell model (cf. Fig. 25), for instance, works best when it is read as being densely composed, since this fits the biological descriptions of the target system on this scale of description. Surely, this does not mean that Fig. 25 can only be construed in this way; somebody else may try to backtrack the numerical solutions or push the zoom factor to its limits, ending up with a correspondence of pixels and syntactic characters. However, as I said, this can only be judged as adequate or inadequate against the background of particular research agendas or purposes. On that score, I am convinced that this is where the methodological level comes into play, namely in support of the choice and justification of individual readings. Consequently, the exemplary classifications (cf. Tab. 1) are not only justified against different rules of combination and interpretation, but also against the functional programs, which have been analysed above (cf. 3.1f.) and which thus form the third, context-variant part of the regulatory framework. The latter is apparently also needed when it comes to defending the attribution of language- and picture-likeness. Although this might be a somewhat puzzling result, the classification could indeed look different in other contexts – even when taking into account the same figures.

This leads us now to the synthesis of the two parts of the framework that are discussed and modified in this thesis, namely the *epistemological* and the *semiotic* part. As for the latter part, this section has actually produced more than "a classificatory result: the semiotic features on which that taxonomy [cf. e.g. Tab. 1; original version: cf. Perini 2005, p. 271] is based do substantive conceptual work by clarifying differences in the kinds of content different types of figures convey" (Perini 2012, p. 146). This is true: recall the example of the unstructured clock-face, where the composition has a clear, limiting impact on the conveyed information (resp. on the accuracy of interpretation). Further examples are easily found; recall, for instance, that Fig. 13 has been graphically refined in order to meet the requirements of publication (cf. 3.1.6).

Accordingly, it is now possible to align the semiotic and the epistemic conditions. On this basis, it should finally be possible to justify an affirmative answer to the question as to whether picture-like systems have the capacity to convey claims (resp. premises) or not, and whether they can thus be integral parts of arguments (cf. 4.2.1; 4.2.3). Further remarks on the supporting conceptual apparatus and especially the semiotic taxonomy follow later, when it comes to revising and enhancing the framework (cf. esp. 4.2.5.1).

4.2.3 A case for the epistemic weight of picture-like means

Having defended an account of the nature of visual means and the conditions of having epistemic weight, respectively, we can now extract relevant implications for the purpose of framing an answer to the question about the epistemic weight of visual (esp. pictorial) means. Here, the answer is still strongly modelled after Perini 2005.

According to the invoked *semiotic* apparatus, visual means can be differentiated into *(written-)language-like* and *picture-like systems*. There are also hybrid systems, which are assigned to *one* of the latter types, being thus either *rather* language- or picture-like.

Language-likeness was associated with the features of being *syntactically discrete* and *articulate*. In order to also obtain *semantic articulateness*, a system is needed where the symbols and their referents are well-distinguished classes, coupled according to pre-fixed rules (as in predicate logic, where a formal language is used; cf. Perini 2005, p. 271). *Picture-likeness*, in turn, was defined in terms of *density* and *inarticulateness* in *syntactic* respects. Both of these features have an impact on the *semantic* level. Namely, even if the system is *articulate* (which is not necessarily the case), it can be hard (if not impossible) to discern its functional components and hence to pair up the symbols and referents.

Now, recall the conditions of having *epistemic weight*: epistemic weight was analysed in terms of the capacity of bearing truth in a formal sense, which, in turn, was seen as being required in order to become part of an argument. The corresponding conditions have been specified as follows: An entity (esp. a supposed symbol) is capable of bearing truth, if it allows for "showing that the state of affairs the symbol refers to is [...] defined at the level of the symbol system [semantic condition]" (Perini 2005, p. 276). In addition, it must "allow for the recursive definition of complex characters from atomic parts [syntactic condition]" (Perini 2005, p. 276; see above).

Casuistry has then revealed that there are (among other sorts) visual representations, (1) which employ spatial (i.e. geometrical, topological or graphical) variables, (2) which are densely composed (at least in parts), so that between each pair of syntactic bricks, there is another one between them, (3) but which are also semantically disjoint and articulate in the way that a symbol refers to exactly one well-differentiated referent (cf. Tab. 1). In our case, this set of features applies to graphs (cf. e.g. Fig. 13): a rather picture-like means with the appeal of a *partially dense, but ultimately formal system*.

However, does Fig. 13 thus really meet the conditions of having epistemic weight *and* being (rather) picture-like at the same time? On the one hand, one could argue that this figure (resp. its suggested reading) fulfils the conditions of having epistemic weight in the way that it is semantically articulate (in contrast to Fig. 25, for instance, which seems not capable of guaranteeing a set of distinct pairings between symbols and referents). Accordingly, it would fulfil the semantic condition, which says that each state of affairs (here: each numerical solution) must be represented by one, and *only* one, atomic part (here: a certain point of the curve), which is predefined at the level of the symbol system (here: by the respective mathematical functions). On the other hand, the figure also exposes syntactic discreteness and inarticulateness, especially as regards the blue and green curves, which line up numerical solutions like pearls of a necklace, making them indistinguishable from another. The problem, which defies a straightforward answer, thus comes with the *syntactic* requirements underlying the condition of having epistemic weight.

In this context, Perini (2005) argues as follows (cf. p. 278): the syntactic parts of a graphical system, especially the lines, are not only related to the semantic class of a *curve*, but – from the methodological point of view *solely* – to certain values which are specific for each point, as each of them "can be given a [*one*] mathematical expression [...]" (Perini 2005, p. 278). This implies that for each of the instances, there is one and only one name – in terms of a certain mathematical function – in the assigned formal language. This, in turn, allows for saying that, even if it is not possible to clearly spot the individual points at this zoom factor, it is still possible to define and program formal truth conditions, which a point must meet in order to fit the rules of the system; namely it must represent a definite value, which in turn can be expressed in terms of a definite location between the axes. In conclusion, "[b]ecause both the symbols and their referents can be assigned linguistic representations, a systematic definition of the concept of truth is possible" (Perini 2005, p. 278). As a result, so Perini's argument goes, the graph (in this sense) is indeed both *rather picture-like* and *capable of bearing truth in a formal sense.*

The question now is whether this answer to the first question of this thesis is satisfying for the intended scope of validity, and whether it can be modified in such a way that it allows also for other, related scopes (e.g. microscopy-like images or other dense systems).

4.2.4 Perini's account: problems and gaps

Let us now step back and discuss the pros and cons of Perini's affirmative account of the epistemic weight of pictorial symbol systems (cf. 4.2.1ff.).[82]

82 Pichler (2010) distinguishes between philosophical problems that concern the generic parts of a frame-work (i.e. the modelling assumptions, if you will) and problems that concern its application (esp. the arguments drawn from or justified with reference to it) (cf. Pichler 2010, p. 253). While solutions towards problems of the first kind are

First of all, compared to earlier, more restricted frameworks (cf. 2.1) one would indeed expect that at least one of the key concepts (epistemic weight or picture-likeness, or both) needs to be modified in its extension, in order to allow for a positive answer. And, in fact, recent literature (not only Perini 2005) proves this to be the dominant strategy. While some obliterate the differences between linguistic and pictorial symbol systems, in order to bring them nearer to each other from the conceptual point of view (cf. e.g. Rudwick 1976; cf. 4.1.2), others tackle the issue from the other end, so to speak, and provide a less rigid account of epistemic relevance (cf. e.g. Rheinberger 1992, 2005). By contrast, Perini (2005) suggests combining a more open account of *argumentation* (leaving out the aspect of subsumption) and a slightly under-differentiated account of *pictorial systems* (which I identified as *rather pictorial*, and which others even classify as *neither pictorial nor linguistic*, cf. e.g. Fischer 1997, yet to which she plainly refers to as *pictorial*; cf. Perini 2005, p. 271). Considering that each of the frameworks thus builds on two major variables, critics have two options to intervene in conceptual terms.

W. Goodwin (2009) and P. Kirschenmann (2010), for instance, tackle Perini's notion of epistemic relevance, which, according to them, is heavily misguided. I must say, though, that this impression is itself due to a misguided reading of her account, which crept into both of their replies. Kirschenmann, for example, writes:

> "[H]aving somewhat misleadingly asserted that [...] visual representations 'must' be capable of being true or false, she [Perini] rather independently addresses the question whether they can be true or false" (Kirschenmann 2010, p. 152).

Consider also Goodwin (2009) in this context:

> "By parallel reasoning, one could move from the empirical observation that numerals are important contributors to the explanations and arguments of mathematicians to the conclusion that numerals must be capable of being true or false. Numerals are of course signs for numbers. Neither a numeral nor the number that it expresses could be said to bear truth. This would be a category mistake. Numbers are objects and objects cannot be either true or false" (Goodwin 2009, p. 375).

This leads both authors to a range of objections, stressing, among other points, that not every part of an argument must be true for itself. Of course, not many words

---

a matter of adjustment to empirically or ideologically justified scopes of recognition, problems of the second kind are of a logical nature, as they demand for an accurate application (resp. fitting) of the acknowledged predefinitions. Both types will be considered in this section.

are necessary to reveal that they are right in this. However, compare this to Perini's (2005) phrasing:

> "To find out if there is something about the visual representations in science that is incompatible with the capacity to bear truth, the first thing to check is whether the features that distinguish visual representations from text and mathematical representations are compatible with the capacity to bear truth" (Perini 2005, p. 263).

The mistake is obvious: both Kirschenmann and Goodwin do not grasp the subtle, but crucial difference between 'bearing truth' and 'being true' (cf. 4.2.1.3). Again, Perini does *not* aim to modify the scope of instances that can be judged 'true' or 'false'. Instead, she examines the semiotic conditions under which the establishment of formal systems is supported and, thus, the definition of formal conditions of truth (cf. 4.2.1.3). Therefore, this objection by Goodwin (2009) and Kirschenmann (2010) can be rejected, for it does not even build on a proper reconstruction of Perini's argument.

Yet, Kirschenmann (2010) adds further doubts, this time concerning the significance of the concept of truth, which is used to specify the conditions of epistemic weight:

> "[I]n analogy with Alfred Tarski's definition of truth, she [Perini] articulates concepts of truth for visual representation. I do not think she succeeds in this [...] endeavour. The references of Tarski's definition are only formalized languages. Correspondingly, the counterparts she comes up with can only support verdicts like 'it is true that in this representation two lines intersect'. Though Perini does not seem to be aware of this, they do not involve any reference to observations or things in the world" (Kirschenmann 2010, p. 152).

In interim defence of Perini, I would like to argue to the contrary that she is well aware of the implications of a formal concept of truth. To me this seems exactly the reason why she chooses it, and rightly so. Since Tarski has shown that it is not possible to define substantial conditions of truth for statements about extra-linguistic entities, Perini refrains from modelling suchlike usages of symbol systems. This does not mean that validation – or whatever is meant when more attention to the "special reference to the real world" is asked for (cf. Kirschenmann 2010, p. 153) – is not an issue anymore. It is just a *different* problem, not yet relevant to research on the sheer possibility of building an argument by the aid of pictorially conveyed information. In order to model this latter setting, it suffices to devise a framework, which offers appropriable conditions so that it is possible to classify systems as *picture-like* on the one hand, and as *epistemically relevant* on the other hand. All *further* (e.g. scientific) reasoning is left to the philosophy of science in a more narrow, context-specific sense, for instance in terms of best practice studies.

Hence, Kirschenmann's objection concerning the missing relevance of a formal notion of truth is mistaken, as it does not hit the heart of the agenda pursued by Perini (2005). His concerns can only be admitted up to the point that the question is very much whether she really needs to choose a terminological trigger like the term 'truth', which is hardly useful for canalising the associations that are linked up with the rich history of this term. Yet, when taking the effort to analyse the assigned framework, the negative effects of this terminological choice are mitigated. Assuming that Kirschenmann did not skip this duty, it eventually seems that he just misses some further remarks on *another* interesting topic. However, naturally not all interesting issues can be considered at this stage of modelling; the surplus potential of the framework are tackled elsewhere, namely when it comes to further applications (cf. Perini 2005ff.; cf. esp. 4.3).

As regards the *semiotic* assumptions, which form the second part of the framework, critics have not been significantly more successful: they either object that it is difficult to specify the conditions of language-likeness (cf. Goodwin 2009, p. 347), which is not exactly helpful, as it simply motivates further research in this area. Or they insist that the combination of Goodman's and Tarski's frameworks is entirely *a-historical* in the sense that it is incapable of capturing the variations that naturally coin the usages of signs and symbols (cf. Kirschenmann 2010, p. 153). The latter objection may sound as follows:

> "I think that meaning, reference and possible truth or falsity of scientific representations equally depends on particularities of their context of presentation and use" (Kirschenmann 2010, p. 153).

There are many variations of this objection which holds that the implementation and use of symbol systems depends on spatiotemporally defined circumstances, which themselves are prone to changes that are hard to capture by a (comparatively stiff) framework. This objection, not merely to this framework, but to *any* conceptual endeavour, is always successful to a certain degree, as theory naturally entails less information than practice is capable of generating. On the other hand, conceptualisation does indeed only work when testing simplified views of the targeted domains; giving up this task means giving up research on efficient frameworks. What is more, in order to forestall this objection, I emphasise a third conceptual dimension, namely the methodological level (also called the level of *picture-based* measures; cf. 3.3.1), where modes and manners of en- or decoding are justified variably, depending on the functional program pursued. This should keep the framework adaptable to historical variances (cf. 4.2.2.2ff.). Yet, I would like to stress that this is not supposed to be a correction of Goodman (1976) and Perini (2005), but rather an explication of a more or less implicit ingredient which underlies their accounts, insofar as both would agree that the utilisation of pictorial means (e.g. in the functional program of representation) varies a lot and, thus, does not necessarily belong to its invariant traits (cf. Bach 1970, p. 119; cf. Goodman 1976, p. 170ff.; also Seja 2009, p. 12; Gabriel 2000).

I think the major drawback of Perini's approach is a different one. In my eyes, it is located on the level of applications, where the otherwise helpful framework reveals some crucial deficits. That is to say, even though Perini's framework is coherent in itself (if you compare her self-set rules and targets of composition and justification), it does not nearly meet the latest expectations of a satisfactory account of the epistemic weight of picture-like means. The blind eye, if you will, comes with the idea that pictorial symbol systems must at least be *translatable* into linguistic symbol systems, in order to obtain epistemic weight. This is used as a supplementary condition, in order to get graphical systems into the equation (cf. 4.2.3). However, as Perini rightly admits, this does not work with all dense systems, but only with those that are embedded in a predefined, semantically disjoint and articulate framing, like Fig. 13, where each atomic part of the curve is distinctly defined by a certain function. As Perini puts it:

> "Some pictorial systems, like graphs, can support a definition of truth because their characters can be systematically named and their content represented with linguistic expressions. This does not seem to work for other pictorial systems, like micrographs. Under the assumption that the characters in the micrograph system cannot be systematically translated into a linguistic system, it is impossible to generate a Tarski-style definition of the concept of truth due to the inability to systematically correlate a linguistic name with a linguistic expression of content of each character in the system" (Perini 2005, p. 282ff.).

Consequently, the attribution of epistemic weight can solely be justified when it comes to systems that are already linked up (e.g. in the mode of illustration; cf. 3.3.3) with a language-like referent (e.g. in forms of a set of equations and definitions), but the account does not help any further when this link gets complex, blurred or entirely lost (e.g. when it comes to as yet undescribed data sets, which are a basic part of the daily work in scientific visualisation, esp. also when the visualisation of data or the analysis of visualised data is delegated to other institutions; cf. 2.3.3, cf. 3.2). Another difficulty is that not every translatable system makes a part of an argument (cf. Fig. 13, which is only an unpublished pre-version that renders interim results more consumable). Moreover, from a similarly empirical point of view, one could add that there is, indeed, also the converse case, namely that not every argument draws from articulate, but also from inarticulate systems (cf. close-up analysis of Fig. 25 and Fig. 26). These systems are thus either not actually picture-like for some reason, which I excluded as a possibility (even for computer-based systems, which some subsume to the scope of 'digital' [i.e. discrete] systems; cf. 4.2.2.4), or one will eventually have to extend the framework in order to also integrate those dense and inarticulate systems. Nevertheless, up to this point, and even in the face of these consequences, one could claim that Perini reaches at least her self-set goals, since what is needed is a modification of her account by means of an *extension* rather than *correction*.

However, the list of objections goes on. I would like to add that the worst problem comes with the fact that Perini is not able to extend the scope of practices that Goodman (1976) *himself* was capable of illuminating, if you consider his notes on graphs (resp. diagrams):

> "What matters with a diagram, as with the face of an instrument, is how we are to read it. For example, if figures on a barogram or seismogram indicate certain point the curve passes through, yet every point on the curve is a character with its own denotation, the diagram is purely analog, or *graphic* [densely composed]. But if the curve on a chart showing annual car production over a decade merely joins the several numbered points to emphasise the trend, the intermediate points on the curve are not characters of the scheme, and the diagram is purely digital [discretely composed]" (Goodman 1976, p. 170).

Apparently, this entails an idea that also enriches Perini's approach, namely the idea that each of the densely arranged points of a graph can be read as a syntactic character and assigned to one, and only one, formal name (cf. 4.2.3). On this assumption, Perini argues that a graph can, indeed, be part of an argument (cf. 4.2). This argument is certainly an add-on. However, I dare say that her approach still comes *so* close to Goodman (1976) that it turns out to be hardly more than a remake. So even though Perini adds Tarski-inspired underpinnings in order to establish an epistemological perspective, she finds herself forced to leave open one of the most challenging tasks of recent philosophy of scientific visualisation: the integration of more complex (esp. 'dense') systems, which are at the centre of interest in philosophy as well as in the examined scientific contexts, if you admit the following report from practice:

> "It is necessary to stress that in their criticism of images, my informants had in mind visualisations of complex domains. They never wanted to extend their criticisms to graphs due to the relatively unambiguous conventional procedures to convert a graph into a table of numbers" (Spencer 2012, p. 36).

Hence, to sum up: On the one hand, Perini's framework supports affirmative answers to the question about the epistemic relevance of picture-like systems, namely when it comes to so-called *graphs*. On the other hand, there is a broad scope of practices which cannot be illuminated when one is merely equipped with the given conceptual apparatus. This does not render her framework obsolete, but feeble. However, since there are no reasons offered for why the integration of such practices should be impossible, and since the state of research offers no alternatives, either (rather a call for research in this area), I will assume this task in the final section of this chapter (cf. 4.2.6; further revisions follow; cf. 4.3).

4.2.5 A case for the epistemic weight of syntactically dense systems

Let me repeat the major drawback of Perini's account and the idea of improvement that goes with it: the analysis has shown that Perini's account offers *no* resources in support of claims about the epistemic weight of specific types of picture-like systems, whereas other types of picture-like systems are successfully accounted for – especially for reasons of closeness to linguistic systems. Yet, at the same time, there is the independent but stable notion that there is a whole inventory of cases indicating that as yet unintegrated systems (esp. syntactically dense and semantically non- or not very well-differentiated systems) are actually being used as integral parts of arguments. In Perini's words:

> "[S]cientific arguments are not limited to verbal and mathematical expressions. More often than not, visual representations are involved. A typical volume of *Science* includes graphs, charts, diagrams, and images produced by various techniques such as electron microscopy, PET scans, and X-rays" (Perini 2005, p. 262).

In our case, for instance, Fig. 25 is intended to mimic the blurry appeal of confocal laser scanning micrographs. Hence, when taking into account the reference measure, it is clear that the figure is read as syntactically dense and inarticulate, for what matters is that the pixels contribute *entirely* and *indistinctly* to the composition of a crowded cell. Also, Fig. 25 is undefined enough, semantically speaking, to allow for only the application of vague classifiers (cf. close-up analysis of Fig. 25). Then again, Fig. 25 is used to "corroborate the claim that there is an optically discernible difference between the considered transduction mechanisms, namely between diffusion (left) and active motor transport (centre)" (cf. close-up analysis of Fig. 25). Even though this claim is to be validated elsewhere (esp. in experimental settings), one cannot belie the impression that it already feeds certain arguments (e.g. about the proper design of experiments). If so, such arguments are not restricted to language- or *rather* picture-like systems.

The problem now is to preserve the integrity of the adopted framework and collected intuitions in the face of equally acknowledged, but not yet assimilated (in this sense *exceptional*) reference cases. This, apparently, has the appeal of a puzzle-solving task (cf. Th. Kuhn 1998, p. 35ff.), as it calls for supplementary analyses, while trying to maintain some of the basic parts of the adopted conceptual apparatus.

4.2.5.1 Continuous vs. strict dichotomies

I shall first of all defend the adopted dichotomies between picture-like and language-like systems on the one hand, and between epistemically relevant and irrelevant means on the other hand, and argue why it is important to keep them *strict*, respectively.[83]

Hence, even though I felt the need to further distinguish between *rather* language-like and *rather* picture-like systems, to begin with, this does not imply that a *continuum* of related phenomena is assumed. On the contrary, it seems unhelpful, if not inadequate, to fuse the categories of language- and picture-likeness when it is also assumed that the defining syntactic and semantic features, on which the dichotomy is built, are *exclusive*. In other words: if certain features (e.g. disjointedness vs. non-disjointedness or ambiguity vs. non-ambiguity) are defined in terms of the absence of the other, they will by definition never characterise one and the same entity; and from this it follows that those features can be *definite* indicators of language- or picture-like systems. Syntactic density, for instance, is considered a defining trait of picture-like systems, while syntactic discreteness (i.e. the feature that is formally opposed to syntactic density) is considered as a defining trait of language-like systems (cf. Goodman 1976, p. 160; Perini 2005, p. 271). It is possible, though, that those features, which are *not* mutually exclusive, occur in synchronisation (e.g. disjointedness and ambiguity), in such a way that a system exhibits either more of the language- or picture-like traits, making it *rather*, yet *definitely*, language- or picture-like.

---

83 P. Koch (1997) distinguishes between 'strict' and 'continuous' dichotomies. For instance, he classifies the dichotomy between *graphical* and *phonetic* systems as *strict*, which is to say that systems are *either* graphical *or* phonetic, but never both at the same time. On the other hand, the dichotomy between communicative proximity ("*kommunikative Nähe*") and distance ("*kommunikative Distanz*") would be a *continuous* dichotomy, meaning that it is possible to find *infinitely many sub-distinctions* in-between the two extremes (cf. Koch 1997, p. 43f.).

In order to keep this account sound and coherent, one needs to introduce a *hierarchy* between the defining features (which Perini does only implicitly), in order to prevent a system expressing the same amount of *definite* language-like and picture-like traits.

Take ambiguity, for instance: ambiguity is not welcomed as a feature when it comes to formal systems. Hence, *ideally* it never characterises a formal system, such as logical calculus. Yet, ambiguity is still a feature that *can* apply to *both* language-like and picture-like systems (see below; cf. 4.3.2.1). Ambiguity is therefore not a compelling indicator of language- or picture-likeness, although some conceive of it as typical of pictorial systems.

By contrast, syntactic density *is* a compelling feature of pictorial systems. That is, a dense system remains picture-like, even if there are further measures of ambiguity reduction. The same holds for the counter case: In Fig. 24, some of the graphical variables (esp. the colours) are not explicitly defined. Still, we find each element distinctly detached from the others against a non-defined background, which keeps the system discrete – thus language-like – despite its (entirely harmless) ambiguities. Even if one takes the spatial arrangement and graphical design of the syntactic components as a (though not equally compelling) sign of picture-likeness, one could admit that the system is *rather*, yet *definitely* language-like on that score. Hence, a discrete system remains language-like, even if the semantic referents are ill defined, and even if the syntactic components are graphically refined and spatially organised. Conversely, a picture-like system does not change in the course of an addition or reduction of ambiguities, or after a linear arrangement of its syntactic elements; we will still decode the whole complex as picture-like, *if* the relation between at least some of its syntactic characters can be classified as dense.

Further examples can be added for other defining traits: In the context under review, Fig. 13 is read as being syntactically dense and inarticulate, but semantically articulate; thus it is only rather picture-like, if one admits that it can be translated into a language-like system. This in turn means that the category of translatability is not as significant as the category of discreteness, but it still has an impact on the classification. It may also be that a figure is read as being syntactically and semantically articulate, albeit spatially organised (cf. Fig. 9; cf. Fig. 24), making it rather language-like. Spatial arrangement is hence also not a fully compelling yet still influential factor. Finally, some systems can be syntactically and semantically inarticulate, spatially arranged and not straightforwardly translatable; they are thus picture-like in the strongest possible sense (cf. Fig. 19 [left]).

To sum up: the strategy of classification – the taxonomy, if you will – as I would like to adopt it from Perini (2005) and Goodman (1976), comes out as illustrated below (cf. Fig. 34). As announced earlier, the taxonomy is already equipped with conceptual supplements, namely in terms of the categories 'rather' language- and 'rather' picture-like, which Perini skips, but which I consider as implicit ingredients of her account that help to mark out its main clue – namely the introduction of translatability – which helps her to align the epistemological and semiotic conditions of epistemic weight (cf. 4.2.1.3).

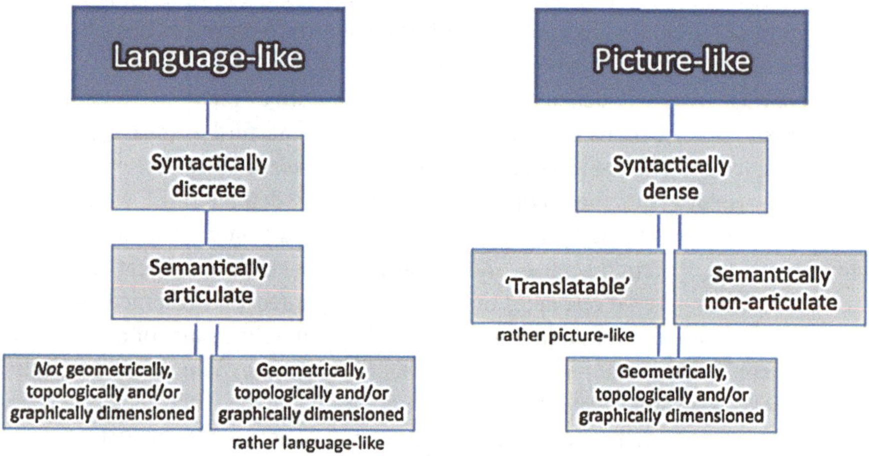

Fig. 34: Taxonomy of the conditions of language- and picture-likeness

Fig. 34 presents the taxonomy on which Perini's framework is built, according to my reading of it. As said above, the taxonomy helps to draw attention to the key modules of Perini's argument, especially to the fact that 'translatability' is (still) seen as an indicator of picture-likeness. At the same time, the latter aspect forms a bridge to the left column, as it guarantees a formal mode of conveying information, which is often seen as typical of language-like systems (as was elucidated above and elsewhere; cf. 4.2.1.3; 2.3).

What is more, the list is also *open*. That is, it is at least theoretically possible to add further features, such as the dichotomy between serially and spatially arranged systems, like Perini does, for instance (following [yet not intentionally] the tradition of Lessing; cf. 4.2.2.1), if you recall that Goodman (1976) does not work with this distinction.

Instead, or in addition, one could also include the feature of ambiguity, which is often seen as being typical of picture-like systems. However, I refrain from assigning it to one of the two sides, since I do not think that there is a clear rationale for that. It rather seems to me that one cannot establish a clear dichotomy in this case (see below; cf. 4.3.2.1).[84] Now, if you subscribe to this framework – bearing in mind that the different 'features' or 'traits' characterise ways of *establishing* or *approaching* the relations between the marks, and not the marks (resp. their relations) themselves (which is why it makes sense to add a methodological perspective to the account; cf. e.g. 2.2.2.2) – the framework eventually fits the definition of a strict rather than of a continuous dichotomy better, because there is no clear distinction in continuous dichotomies. This means hybrids are the standard cases that one needs to care about, for instance by defining the extremes between which they oscillate. By contrast, in a strict dichotomy, such as the one presented here (cf. Fig. 34), real hybrids obtain the status of exceptions or sheer impossibilities.

This also holds for the dichotomy between epistemically weighty and non-weighty means, which builds the second part of the framework. This dichotomy needs to be considered as strict, too. This is also why it can be linked to the semiotic framework.

In fact, according to Perini (2005), the conditions of epistemic weight correspond to the semiotic conditions that establish a formal system, namely the conditions of being finitely differentiable in syntactic and semantic terms (cf. 4.2.1.3). The latter conditions leave no room for an in-between state of being more or less epistemically weighty, since either a system expresses these traits, or not (cf. Tab. 1). It might just be that a system expresses the syntactic yet not the semantic condition, or, conversely, that it expresses the semantic yet not the syntactic condition, as in the case of graphs, which are considered as having epistemic weight on that basis (in contrast to earlier accounts; cf. 2.3.2, 2.3.3). This is Perini's key proposal in response to the question about the epistemic weight of pictorial means, together with the notion of translatability, which creates a bridge to epistemology, and the revival of the tradition-rich dichotomy between spatially and serially composed systems, which supports seeing graphs as picture-like.

Thus, all I would like to stress up to here is that it is neither necessary nor helpful to deviate from the idea of strict dichotomies. I would rather like to point out that it helps to explicate this formerly implicit assumption, not only since it guarantees an adequate construal of the reference account, but also since adopting

84 Since the analysis of the epistemic surpluses (here esp. 4.3.2.1) helps to complete the taxonomy above (resp. its justification), the two major problems of this thesis eventually turn out to be interlinked, since their treatment yields complementary aspects for the other problem, respectively.

the idea of strict dichotomies means taking the question on the epistemic weight of picture-like systems seriously as an *either-or*-question. Thus, there will be no escape into a continuous view in the following sections.

4.2.5.2 The concept of ad hoc syntaxes

The literature survey (cf. 4.1), and particularly the review of Perini's account (cf. 4.2.3), has shown that there are crucial deficits when it comes to integrating positive intuitions on the epistemic weight of genuinely picture-like systems. The integration is hindered from the viewpoint of the proposed conceptual apparatus where the notion of picture-likeness is not appropriable to, if not conflicting with, the notion of having epistemic weight, and vice versa (cf. 4.2.4). In other words, if a system is taken to be densely composed, which is the condition of picture-likeness, it is not taken to be well-differentiated, which however is the complementary requirement of having epistemic weight. As a result, the question concerning the epistemic weight of pictorial systems must either be rejected or left open – even in cases as significant as the use of microscopy-like images (cf. close-up analysis of Fig. 25 and Fig. 26; see also below) – which apparently collides with scientific practice. And so, again, what we are faced with is after all a problem of systematic integration.

However, I claim that by adding the concept of *ad hoc syntaxes*, the already modified framework will also allow for the epistemic weight of exactly such – as yet unaccounted for – systems, namely systems which are approached as being syntactically dense, semantically non-articulate and spatially dimensioned (cf. 4.2.5.1). For this to be achieved, the concept of ad hoc syntaxes must allow for the redefinition of pictorial systems in a way that preserves their integrity as picture-like entities and nevertheless makes them classifiable and adjustable as to the conditions of having epistemic weight.

Definition

In the choice of terms, the phrase 'ad hoc syntax' evokes two associations that help profile the concept so denoted. First of all, by the choice of the term 'syntax' the phrase appeals to the *syntactic* level of establishing or approaching symbol systems (cf. 4.2.2.4). Classically, this is where it is decided (ever since Goodman 1976) whether a system is language- or picture-like, if you recall the suggestion and appraisal of syntactic density as the defining trait of picture-likeness (cf. 4.2.5.1). Secondly, in the choice of the term 'ad hoc', the phrase appeals to trial- and prudence-based – in any case unscripted, case-orientated and notoriously tentative – accounts of that entity's instances to which the attribute is assigned (i.e. the syntax). Taking these two parts together, the phrase denotes a specific account of the syntax of a symbol system. The account is specific in the way that it test runs one or more options of characterising a system in syntactic terms. This is also why it is both semiotically and epistemologically dimensioned; it relates to the *syntax* of a system, yet in the mode of *exploring* relations. This surely requires systems that are not yet sufficiently prefixed in that sense, but still appropriate enough to be worked with (see also below; cf. 4.2.5.4).

Further aspects – less specific, yet still significant – are that such accounts may depend on options of intervention (cf. 4.3.3.3), since some systems are so complex that ad hoc syntaxes must be marked interactively in order for the user to be able to assimilate the test run implementations, depending on individual skills and needs. In this context, graphical *feedback systems* may become indispensible and in this sense also typical of this kind of approach. Hence, depending on the complexity of a system, the users may rely on optical markers which enable them to trace the test run specifications of a system's syntax.

But before getting into the details, I first need to indicate why this solves the problem at stake, namely, to show that the proffered semiotic and epistemic requirements are satisfied by the practices so described. First of all, are systems that are specified in this way still picture-like? I claim that they are, for ad hoc syntaxes render them tractable via *test-run* specification. With this, however, the indefinite spectrum of options of specification and thus the defining aspect of picture-likeness is maintained as a logical part of the ad hoc treatment (this point is further stressed below; cf. 4.2.5.3). Moreover, the question is, whether ad hoc-specified systems meet the epistemic condition of being truth-apt in a formal sense. My argument in support of this is that there can be islands of specification in the context of dense annotation. In some cases, these 'islands' are tentatively demarcated and thus rendered formally tractable by the aid of auxiliary signifiers. The point that still rescues formality is that as long as there is a well-specified coupling of syntactic and semantic entities – like 'blurry scuffs are to be taken as traces of transported proteins' (cf. close-up analysis of Fig. 25) – neither the tentative mode nor the limited scope of such a specification affects the formal integration of these semiotic parcels (cf. esp. 4.2.5.4).

Certainly, these are only the first hints of the main line of argumentation associated with the introduction of ad hoc syntaxes, which is still in need of substantiation. Therefore, I will now review the examples that have been introduced earlier (cf. 3.1f.) in order to add more empirical detail and pave the path for a further degree of precision.

Examples I

The so-called 'microscope-like images' (which I prefer to call 'microsco*py*-like images'; cf. e.g. Fig. 25 [a], [b], Fig. 26 [a], [b], [c]) should help to illustrate the functional role of ad hoc syntaxes (as promised on several occasions in this thesis) and thus to flesh out the identikit picture, which has been drawn by the – still rather abstract – definition above. Focusing on microscopy-like images also has the advantage that – as far as I know – they have not yet been documented in literature so far.

Microscopy-like images are, first of all, used in contexts, where the scope of semantic referents is predefined. There maybe, for instance, elaborate notions of a cytoskeleton, signalling molecules, etc. (cf. 3.2.10). Yet, these semantic predefinitions don't seem to be fully binding, if you recall that there are alternative renditions, which represent the same referents more distinctly, yet which are not favoured on that score (i.e. geometrical renditions [cf. 3.2.4ff.] and graphs [cf. Fig. 25 [c]; Fig. 26 [d]). Therefore, on the one hand, there is a pre-fixed inventory of semantic classes, which brings the systems close to being formal ones. On the other hand, this inventory is not articulately instantiated by the microscopy-like images, compared to other renditions. On the contrary, the microscopy-like images are *rendered inarticulate* in this syntactic sense, for it has become impossible to clearly distinguish the syntactic elements that couple up with the semantic instances, and that can be variously combined in order to form more complex syntactic classes (this is the syntactic condition of formal systems; cf. 4.2.1.3). The images are thus transposed into a state of vagueness, which also characterises micrographs that *are* truly generated via confocal laser scanning microscopy (cf. Fig. 19 [left]). Microscopy-like images are thus *blurred on purpose*, for they are meant to represent initial versions rather than fixed accounts of the referent system. This has the advantage (methodologically speaking) that the images can be compared with the pictorial upshots that are generated at corresponding stages of empirical research, namely the stages of *data acquisition* and *unfinished data analysis*. From the similarities and disparities between the pictorial items, researchers may then establish whether the model systems expose the focal traits of the target system as modelled in experimental settings (cf. 3.2.10ff.). However, in line with the main focus, insights are only expected in syntactic terms. Thus, there will be no discovery of new cellular ingredients, but perhaps of untried relations amongst the modelled ones.

Then again, the surfacing of relations (epistemologically, not ontologically speaking) is not yet sufficient reason for the assignment of epistemic weight to the examined systems (according to the conditions of epistemic weight set up before;

cf. 4.2.1.3). The condition for this assignment should rather be to (re-)establish the systems so that they allow a formally strict coupling of syntactic classes and semantic referents. This would certainly be easy, if it were just a matter of removing some previously added graphical effects. Yet, it gets tricky as soon as large data sets are more or less automatically brought into shape (cf. 2.3.3). In such cases, the establishment of the syntax is partially delegated to detecting and calculating devices and hence becomes a matter of *re*-cognition afterwards.[85]

However, such cases, where the compositional traits of a rendition rely on automated inputs, are all but rare in current research (cf. 2.3.3), and in such cases, the strategy is to determine areas of interest, which can be referred to tentatively as syntactic classes. Defining such areas means narrowing down the scope of shapes, locations and relations of the syntactic components of a system (e.g. areas where the signalling molecules cluster up or where they come across while moving; cf. 3.2.10). This happens, as indicated, in the case of pre-fixed reference bases, including semantic pre-definitions drawn from empirical accounts or elsewhere, the methodology of the picture-generating devices and measures (cf. 3.2.13) and the optically accessible outputs on computer screens (cf. 4.2.5.4). Yet, as stated above, the resulting syntactic profiling is not at all final; it is merely *ad hoc*, following the methodology of trial and error. This is why ad-hoc syntaxes generate expectations rather than convictions and why, concordantly, *the option of change is maintained.*

In our case, for instance, Fig. 25 serves as a reference when it comes to justifying the claim that certain molecules are transported, whereas others diffuse through the cell (cf. 3.2.10). It might then be that, after a closer examination (perhaps aided by a zooming device), certain areas need to be redefined as background parts of the cell, whereas others need to be redefined as areas of interest. Consequently, the amount of clustering molecules may turn out to be bigger or not, or to have moved slightly or not, etc. There might also be an emergence of syntactic classes, such as *clusters* of syntactic elements, which are reminiscent of certain shapes (e.g. sets of molecules may resemble clouds, etc.).[86] In fact, such interplays between the geometrical, topological and graphical descriptions of a system and the syntactic (ad hoc) classifications of the traits described is most promising in heuristic terms, as it may lead to new accounts of the visualised data-base.

---

85 The corresponding workflow, as modelled in scientific visualisation (i.e. the 'visualisation pipeline'), has been introduced in the context of the second literature review (cf. 4.1.1; esp. Fig. 32).

86 The emergence of shapes is most pronouncedly documented in research on fractals.

However, it would be a different case if new or further insights into the target system (e.g. after an adaption of experiments to hypotheses drawn from the exploration of the mathematical model; cf. 2.1.10), call for a full revision of the model, not only of the interpretation of its solutions. This, of course, would concern the level of pre-definitions (here: the assumptions that concern the cellular interior), which need to be allowed for by the model. Yet, this is not what I have in mind here. For now, I just would like to capture the as yet unaccounted phenomenon of syntactic test-run settings, that is, classificatory interplays with what may be referred to as the 'architecture' (i.e. geometrically, topologically or graphically describable traits) of one and the same rendered data set.

Interim summary

To sum up: On the one hand, there are systems semiotically approached, where the challenge is to find or refine the *semantics* (such as photographs of unknown species or landscapes), and where I may therefore ask: *what do the single parts of the image refer to*? On the other hand, there are also systems where the semantics is pre-defined, but the syntax is in the focus of the viewer. This is, for instance, the case with the microscopy-like images under review. Here, one may ask the question: our model is meant to describe a certain system's behaviour (e.g. the diffusion of signalling molecules), but, facing the visualised solution of the model, *where exactly are the corresponding elements and how do they relate to each other*? In both of these cases, the so-called data analysis (which has been analysed in terms of a workflow, e.g., by Wimsatt 1990; cf. 4.2.1) may cause surprises or confusions amongst the users. Yet, a closer look reveals that the respective settings differ from each other, because these reactions are caused either by a lack of specification on the syntactic *or* on the semantic level.[87] Now, the follow-up task is to find further evidence for this distinction and especially for the idea that syntactic test-run systems cannot only somewhat capture to some extent but also *illuminate* the yet unaccounted traits of practice(s). Therefore, the question is still why things make more sense now.

Ostensive Signifiers of ad hoc Syntaxes

So far, it seems that ad-hoc syntaxes are intelligible in the sense that they are set up and apprehended only by the intellect which approaches an optically accessible area in that way. If this were true, there would be no clearly detectable traces of

87 This is just another hint to the notion that computer-based systems do not per se obtain or express a dis-crete, ready-to-hand syntax, which I already objected to earlier in this thesis (cf. 4.2.2.4).

ad-hoc syntaxes; the latter would only surface in what is *said* about and *done* with a system and how one thus relates to it discursively as well as by the choice of dedicated operations. However, I claim that there *are* instances of ad hoc syntaxes, which are optically accessible and which can be assigned to that category, for they are meant to reproduce, steer or hinder ad hoc specifications (esp. in the context of publication).

In point of fact, the act of defining an *ad-hoc syntax* can be marked by itself, namely by the aid of signs, which are given an ostensive function. This is to say that those signs are specifically used to link the symbol system with its specifying context. In other words, the signs do not, or do not only, belong to the examined system (e.g. to the visualised solution of a cell model), but they are instances of auxiliary systems, which help to establish the ones to which they are related. The following extract from the empirical reference base (cf. 3.2) may illustrate the point (cf. Fig. 35):

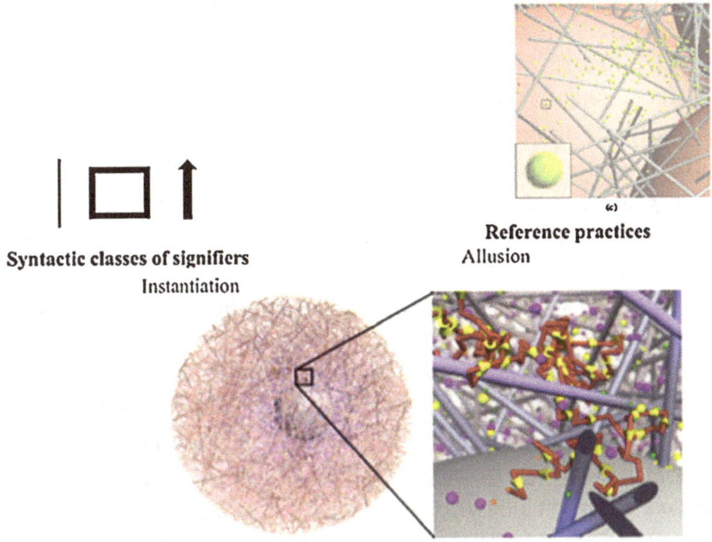

Fig. 35: Signifiers and signification of ad hoc syntaxes

Fig. 35 shows a system that is equipped with signs that guide the ad hoc specification of its internal syntactic organisation. In this case, the cell model is first rendered in such a way that it occurs as a cluster of particles (down left), which is then refined (down right). The two renditions are linked to each other by virtue of a black square and two lines that mark an area of interest in the left figure and tie it to the right figure. The latter thereby comes to appear as a syntactically more refined close-up of the left figure.

To guarantee the effect of guided specification, the signifier needs to be identified as such. Hence, the user needs to relate it – as with other signs – to *types* of signifiers in the mode of *instantiation* (above left). And the user needs to relate the signifying practice to familiar practices in the mode of *allusion* (above right). Allusion is also a type of reference, less rigid than an instantiation of a class or category, as the target of reference is not a catalogue or alphabet of practices, but an open set of *reference cases* (rather than *types*), which follow more or less the same methodology. In the case under study, for example, the allusion to equivalent systems already concerns other figures in the same paper (cf. Fig. 19 and Fig. 27), where black squares are used to draw the attention to areas of interest.

Yet, such ways of linking between practices are neither made explicit, nor given any constraints. Concordantly, one is allowed to take into account other role models (in terms of familiar practices) across times and places. The signifier (resp. its use) thus transcends the present context in at least two ways: firstly by reference to *common types of signifiers*, which, in this case, draw on so-called 'geometric' or 'graphic primitives' (like lines and squares) that are particularly common due to their simple (in the sense of internalised) shapes (cf. Bungartz, Griebel & Zenger 2002; cf. 4.3.3.2), and secondly by reference to *practices of indication or allusion*, which may have a long tradition, and which may (perhaps for this reason) even guide the use of new types (e.g. microscopy-like images).

The targeted measures of indication or allusion may also be conceived of as ways of *pointing to* the proper ad-hoc specification; this is also why I refer to these signs as having an 'ostensive' function. The two black lines in Fig. 35 (resp. Fig. 19), for instance, which reach out to the figure on the right could also be replaced by a pointing finger or the use of a pointing device. In fact, ad-hoc syntaxes may be marked exactly by virtue of such rudimentary acts of *pointing*, which have been well considered in literature so far (esp. in times where pictorial systems were analysed in regard to performative conditions), yet which have never been saliently related to the debates under review.

Examples II

The neglect of ad-hoc syntaxes and ostensive signifiers in previous accounts is quite surprising, as it is not necessary to quarry long for examples. In Perini's paper, for instance, there is an arrow added to a micrograph, which has been drawn from a low temperature electron microscopy (cf. Perini 2005, p. 272). The arrow points towards an area of interest, which is therefore marked as a syntactic test run correlate of an expected or otherwise preconceived semantic class. Still, Perini makes no mention of the arrow here or in any other of her papers; even though the arrow guarantees the efficiency of the system in the context of argumentation, as it specifies an area of interest, which thus meets, at least for the parts relevant to argumentation, the syntactic condition of formal systems.

In our case, we find that signs obtain analogous functions: Fig. 19 employs graphical primitives in order to draw a connection between a distant and a close-up view of an area of interest (see above; cf. Fig. 35). Other, textually more integrated versions of signifiers occur in forms of phrases such as "(cf. Fig. 1 Sec. II)" (cf. Schittler et al. 2010, p. 4), which may (as in this case) build on a colour code in order to draw attention to the area of interest addressed. In other cases, the specification and marking of areas of interest is even technically supported (e.g. by options of interactive rendition), as in case of the cell model under consideration here (for further information on the technical support of interaction; cf. 3.2.12). So after all that, it seems that by introducing of ad-hoc syntaxes and related ideas (e.g. on ostensive signifiers), it becomes, indeed, possible to access and integrate a whole lot of formerly neglected practices.

Summary and Conclusion

The systematic role of the new concept in the adopted framework remains to be specified: I consider it to be another bridging principle, complementary to that of the translatability of systems (cf. Perini 2005, p. 278). The latter was introduced to allow for the epistemic weight of symbol systems that expose a syntactically dense and inarticulate design, but well-differentiated semantics. In Perini's case, this amounted to the integration of graphs into the array of epistemically weighty systems (esp. densely composed curves in coordinate systems). In contrast, the concept of ad-hoc syntaxes allows for a *syntax-focused* extension of the framework. It helps explain that and how syntactically dense *and* semantically inarticulate systems are rendered *tentatively well-differentiated* by the aid of auxiliary systems. Thanks to the analysis of the latter, it is possible to allow for the epistemic weight of genuinely picture-like systems, which has only been backed by affirmative intuitions so far (the explanatory merits are elucidated once more below; cf. 4.2.5.3). This revised view of the epistemic weight of such means would not have been possible without Perini's preparations, but nor does it seem that they were based on them alone.

4.2.5.3 Explanatory potential

In this section, I will take another look at the concept of *ad-hoc syntaxes* – this time with special regards to the contributions that can be drawn from it against the background of current research. Some benefits have already been mentioned above, but they surely deserve their own section, also because they multiply at a second, more focused glance.

First of all, the concept of ad-hoc syntaxes has the unbeatable (already appreciated) advantage of bringing further phenomena into the scope of conceptual reach (cf. Fig. 36). On the one hand, this concerns genuinely picture-like systems, which expose a dense and inarticulate syntax as well as loosely defined, syntactically yet uncoupled semantics (cf. 4.2.5.3). Here, the notion of ad-hoc syntaxes helps to explain how such systems are rendered tractable via ad-hoc specification (resp. approximation), while the indistinct spectrum of *options of specification* – and thus the defining trait of *picture-likeness* – is maintained. On the other hand, this concerns strategies of signifying test-run specifications of a system's syntax. Here, the concept of ad-hoc syntaxes provides for awareness of auxiliary semiotic means, which are neither entirely absorbed into, nor fully detached from the test-run systems, but which essentially help process them.

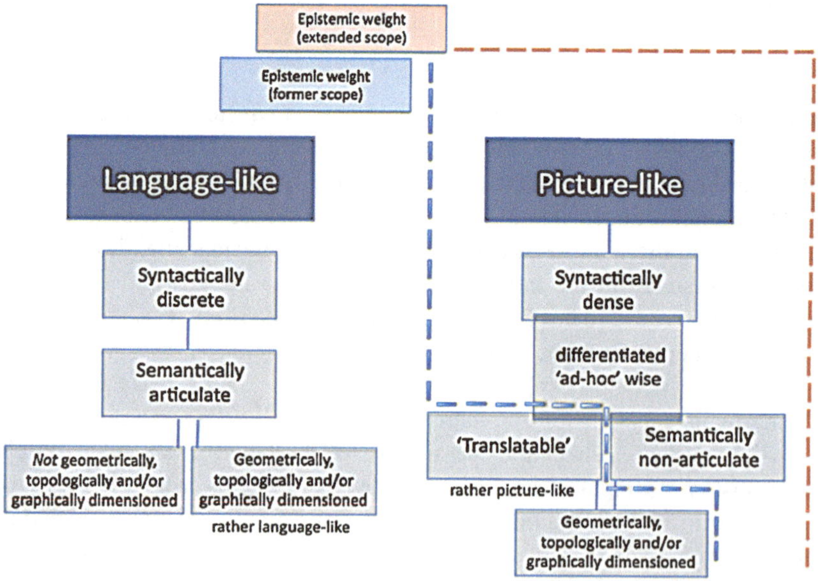

Fig. 36: Extended scope of systematic integration

After all, the concept of ad hoc syntaxes as well as the other concepts that are brought to light in this context, such as the concept of ostensive signifiers, are thus meant to add to the efficiency of formerly introduced bridging principles, like the notion of translatability (cf. Perini 2005), which turned out to be too limited (cf. Fig. 36; cf. 4.2.4).

When considering the explanatory potential of the concept of ad hoc syntaxes, two objections, which could possibly jeopardise the whole idea, come to mind.

First of all, one may insist that there are systems which are conceived of as being both picture-like *and* functionally integral to arguments, but which are neither approximated ad hoc, nor set into relation to any well-defined, formally constrained semantics. In other words, one may question, thanks to the notion of ad-hoc syntaxes, whether, in the end, one has the resources to comment on the epistemic weight of *all* relevant instances. A first objection could accordingly be directed against the claim of *comprehensiveness*.

There are, for instance, entirely monochromatic works of art, which appear to be both *unspecific* and *kept unspecified* in all conceivable respects (i.e. in terms of composition, semantic referents and methodologically justified usages). Yet, they may nevertheless be used to feed arguments. For instance, one may use these works to simply argue that they *are* monochromatic or abstract (or both) in the suggested sense.

However, what appears to be a straightforward objection turns out to be difficult to uphold. One could first of all reply that, when it comes to the classification of pictorial systems, the latter work *in mere support* rather than as *integral parts* of the respective arguments, because it is not the image which is integral to arguments on its own classification, but extracts of its own descriptions (cf. 4.2.1.2). Critics would thus need to find examples where such systems are used to extract data in order to ensure the integrity of an argument. This, however, should be difficult to accomplish if, at the same time, it is claimed that these particular systems are designed to *exclude* such extractions.

Yet, there is even better support than this. In order to show that there *can* be systems, which have the capacity to become integral to arguments, although they do *not* meet the semiotic conditions of formal systems (as specified earlier; cf. 4.2.1.3), one would need to show that arguments do *not* depend on formalised reference systems, that is, systems, which build upon distinct and limited – either predefined or ad hoc specified – couples of syntactic classes and semantic referents. However, I claim that this can hardly be shown, since, even if one tried to pinpoint exceptions, one would profit remarkably from the notion, which has to be rejected, of distinct couples of syntactic and semantic classes. Consider, for instance, an entirely monochromatic image. The latter exposes only one colour in the area between its constituting frames. It would thus seem that it could be left syntactically unspecified, yet it could be still argued, on account of the conveyed information, that the artist aimed for an idealised view of the sky or for an undisturbed note of blue, etc.

However, there is an error involved: if one considers this image, or the hereby conveyed information or the assimilated optical input, as monochromatic, one certainly relies on a *specific* syntactic class, only that the latter is deemed extended over the whole area. In other words, in order to feed arguments by reference to the 'unspecific appeal' of a picture-like system, there *are* test run specifications required, which may, however, yield only one, spatially overarching syntactic class. The latter can also be assigned a referent, depending, for instance, on whether the colour used is considered to be 'figurative' or not, which is a problem in its own right. But even in order to argue that there is no distinct semantic referent, or that the semiotic variable is left entirely empty, one needs to draw on the idea of referents; the semantic part of the couple would hence be negatively yet still *distinctly* specified for the sake of an argument. And even if one tries a weird coupling of marks and referents, claiming, for instance, that the image represents a green carpet seen through a pair of blue lenses, such a construal depends on the idea of distinct couples of syntactic and semantic classes, yet in the mode of rejecting the usual methodology. Therefore, likewise games downright *live* from the principles they confute. And so, I claim that, even if there is no way of proving comprehensiveness strictly and rigidly, it is *easier* to give in into the claimed explanatory potential, than to show that there are exceptions to it in terms of entirely unspecified yet still integral parts of arguments.

The second objection against the explanatory potential of the concept of ad hoc syntaxes is not directed against the comprehensiveness, but against the *consistency* of the related account. To put it bluntly: one may object that there are different ideas of *formal systems* involved, which is why it only *seems* that the account is now sufficiently comprehensive. The question thus is whether ad hoc specified systems *are* formal in the premised sense. However, since this question strikes at the very heart of the account, it is tackled separately (cf. 4.2.5.4). Meanwhile, I deem the major explanatory potential defended.

Yet, this is not all there is to say about the explanatory merits of the extended framework. Besides boosting the argument on the epistemic weight of genuinely pictorial systems, the concept of ad hoc syntaxes also carries *explanatory surpluses*, if you will.

It helps, for instance, to model the inter-relations and dependencies between the different levels of establishing or approaching symbol systems, that is, between the compositional (resp. syntactic), representational (resp. semantic) and methodological levels. The notion of ad-hoc syntaxes has potential in that direction, as it draws focus on *acts of choosing*, namely between different options of specification on the one hand, and different options of signification on the other hand. Such choices make sense only or particularly in the light of the choices *on other levels*; and in this sense, they link these levels, respectively. Consider, for instance, the three-dimensional numerical model presented by Schittler et al. (2010) (cf. 3.2). The choice of three dimensions is not only methodologically justified (i.e. in regard to the cell model), but also with reference to the limited options for encoding *more* dimensions (cf. 3.1.6). Accordingly, the semiotic (esp. compositional) boundary conditions, *including the limits of auxiliary systems*, which apparently *also* fail in guiding the specification of more than three dimensions, have an impact on the scientific agenda. On that note, the notion of ad hoc syntaxes turns out to be a promising point of reference when it comes to strengthening the distinction between the different levels and associated rationales of establishing or approaching symbol systems.

Moreover, insofar as the notion of ad-hoc syntaxes sheds light on measures of choice and attention steering, it also calls to mind the significant role of *interventions*. Interventions (esp. in terms of ad hoc specifications) may surface thanks to the use of *ostensive signifiers*. The latter are applied to systems in order to guide their construal, but they leave room for modifications and variations just as other usages of symbols do. This might be the reason why the authors considered did not particularly distinguish such usages of symbols (resp. of graphical variables in general), although the latter obtain a vital role in justifying the epistemic weight of genuinely picture-like systems, insofar as they guarantee the semiotic conditions of discursive integration. Therefore, on the way to justifying the epistemic weight, the concept of ad hoc syntaxes draws attention to basic practical conditions of the implementation and exploitation of symbol systems and, for this reason alone, I consider it worth adding to the dedicated framework.

Last not least, the concept of ad hoc syntaxes also helps elucidate how symbol systems come to have *borders*. In anticipation of the subsequent section (cf. 4.2.5.4), it is worth mentioning that the establishment of borders is needed in order to talk meaningfully about *formal* systems; since whether or not a system is semantically disjoint and unambiguous – that is, whether or not it holds that each of its symbols refers exactly to one referent in all contexts – depends on measures that guarantee distinct and limited couplings of syntactic and semantic classes and hence also a limited scope of reference contexts. Take Fig. 15, for instance. One could say that colourful curves, and the dots they consist of, are widely used as syntactic elements and probably given different functions in the other contexts. In order to prevent Fig. 15 from becoming non-disjoint and ambiguous on that score, there are different options to choose from. These options share the common characteristic of either establishing or cutting a link to other symbol systems or contexts of use. Guiding associations in that sense can be achieved, for example, by adding citations, which serve as *attractors*, insofar as they draw explicit links, or by *graphical alienations*, which serve as *distractors*, insofar as they cut links or confute attempts of linking. In case of Fig. 15, for instance, there is no link to the meaning of the colours, but the graphical alienation evoked by them, compared to Fig. 14, prevents one associating the figure with the bifurcation diagram presented previously (cf. close-up analysis of Fig. 15). Hybrids are also possible: for instance, a system may be considered 'formal' with respect to the limited set of numerical values that are distinctly represented by exactly one of the densely arranged points. Then again, the very same system may employ the colour red and use it for more than one purpose (i.e. for marking heat *and* danger), which does not make it less formal, if you follow Perini (2005), even though a semantic distinction seems to be confuted by this.

But before I come to review the conditions of formal systems (cf. 4.2.5.4; 4.3.2.3), the important point to reiterate is that the notions of ad hoc specification and ostensive signifiers are worth mentioning, when it comes to explaining the borders of semiotic systems.

After all, a whole lot of issues that casuistry is faced with (e.g. the question how one finds appropriable criteria for judging whether a system is formal or not) can be brought up along with the concept of ad hoc syntaxes and the ostensively used traits that are systematically linked with it. Further explanatory potential of the extended framework will be revealed below (cf. 4.3).

4.2.5.4 Ad hoc syntaxes in formal systems

In the previous sections, the conditions of epistemic weight were analysed in terms of the semiotic conditions of formal systems (cf. esp. 4.2.1.3). Therefore, in order to ensure that the notion of ad hoc syntaxes truly adds to the efficiency of like-minded approaches, the task is to check whether ad hoc specified systems *are* actually formal in the premised sense. This is also important because ad-hoc specified systems appear to lack a trait or aspect, which is often said to characterise formal systems, namely that of being subject to a constraint set of components and rules of formation.

The usual problem is that there are *various* ideas of what an entity must be like, in order for it to be classified as 'formal'. Perini (2005), for instance, draws on Tarski's notion of formal truth, according to which a statement is 'formal' (resp. 'formally true'), if it agrees with the definitions set forth for the formation of statements (cf. 4.2.1.3; cf. Tarski 1956). In the case of semiotic systems, this comes down to rules of composition (syntax) and interpretation (semantics), which set constraints on how the components of semiotic systems must relate to each other in order to build formally accurate composites. For others, however, 'formality' rather comprises conditions such as *minimal equipment* (in terms of a *low* number of symbols and rules of combination and interpretation) and *prescriptive benchmarks* (in terms of a *limited* number of symbols and rules of combination and interpretation) (cf. MacFarlane 2000). Again others conceive of formal systems as relying on symbols that are stripped of a fixed meaning or context of use. On that note, symbols or symbol systems are 'formal' in the sense of *being reduced to their 'form'* (cf. Carnap 1937; Tarski 1995; Krämer 1988; MacFarlane 2000).

Now, as said above, this not even fully exhausted variety might raise the suspicion that one needs to borrow from divergent ideas of 'formal' systems in order to achieve the integration of genuinely pictorial systems into the scope of 'formal' – therefore epistemically weighty – systems. An argument in support of this suspicion could make use of the definitions launched so far; because, given that *ad hoc syntaxes* occur in the context of ill-differentiated symbol systems; and given that – in accordance with Perini (2005) – a symbol system is formal (resp. formally correct), if it is possible to verify recursively that it instantiates a limited set of well-differentiated rules of combination and interpretation, then the ground seems prepared in order to argue that ad hoc syntaxes cannot be part of formal systems. This is obvious then, because the ad hoc specification of an as *yet* ill-differentiated symbol system cannot be an instance of a limited set of well-differentiated rules. In other words, the definition of ad hoc syntaxes can itself not be verified with reference to a limited set of well-differentiated rules; otherwise the entire system would *be* well-differentiated and, thus, not in need of an ad hoc syntax.

To repeat the core of this train of thought: if the concept of ad hoc syntaxes refers to a test-run specification of rules where there are none as yet, then there is apparently no rule or scheme for this. Consequently, the notion of ad hoc syntaxes does not seem to comply with the requirements of formal systems, which is a problem insofar as these conditions were set in correspondence with the conditions of having epistemic weight (cf. 4.2.1.3).

The follow-up question accordingly is why ad-hoc specified systems adhere to the pre-set terms anyway. In this context, I would like to present several arguments in defence of this notion. To this end, I will extend the analysis of *formality* and claim that the result is still compatible with the adopted (already modified) framework. The extended framework is thus only incoherent to the extent that the suggested idea of formal systems lacks a more elaborated analysis and explication in the initial mindset. Taking note of the arguments below, one may therefore come to agree that we need to clarify the idea of formal systems, rather than our intuitions about the epistemic weight of picture-like systems.

The knot opens when recalling the *practical* conditions of formality. These are best revealed in the light of the problem (resp. paradox) of ultimate justification. Accordingly, the following aspect is to be added to the argument above:

An *ultimate* set of well-differentiated rules for the specification of well-differentiated rules requires *itself* as a premise. In other words, the search for an overall formal system that would guide the establishment of other formal systems is indefinite, on account of the fact that it requires a notion of formality where there is none defined in formal terms yet. From this, one may extract the argument that there is no formal system which is *solely* defined with reference to formal systems.

What follows from that? Since formal systems are themselves *not* instances of entirely formal systems, their formation must follow some kind of escape strategy (preferably *test run breakups*) since leaving them with infinite iterations is inconvenient. This is why, particularly in empirical contexts (like the ones under study; cf. III), it makes sense to turn the focus to practices of compensation (e.g. ad hoc specifications), which are intended to keep up the formality of systems where there is *no* ultimate formality possible.

The first argument is a straightforward follow-up: I claim that ad-hoc specified systems may, indeed, be classified formal, because, in point of fact, formal systems can be *encapsulated* in otherwise non-formal systems. Accordingly, I claim that there are *islands of specification*, like certain areas in an electron micrograph (cf. Perini 2005, p. 272), which are marked out (e.g. by the aid of an arrow) as test run instances of distinct couples of syntactic and semantic classes. The use of the signifiers (e.g. arrows) can be justified recursively anyway, namely with reference to the distinct couple of a syntactic class and semantic referent, which is preferably mentioned in close context. Neither the spontaneity, nor the limited scope of this specification (compared to the extension of the semiotic system) does actually affect the formality of this way of approaching or forming the system, since, again, there is a *distinct syntactic class coupled up with exactly one referent in the relevant context* (e.g. of an argument). And so, it turns out that it is important not to mix up the concept of formal systems and the concept of picture-like systems, since the framing of formal systems is not necessarily equivalent to the framing of picture-like systems. Consider Fig. 15, for instance. Here, the use of graphical variables (esp. colours) is quite unspecific. Still, Perini would refer to it as a 'formal' system in the claimed sense, since, in other respects, it is subject to the idea of distinct couples of syntactic and semantic classes. Another example would be Lagrange, who thought of figures as being non-formal in use (cf. Lagrange 1997). Still, these and other declarations are presented to us in natural language, without further ado. Accordingly, even in this case, which is taken from the era of rising formality, we will admit that the non-formal parts do not render the whole treatise non-formal, which leads us to conclude that, in order to fully grasp the explanatory potential of ad hoc syntaxes, it is first of all important to subscribe to the notion *of encapsulated formality*.

In the second argument, I would like to jump on the same bandwagon. The claim is that neither the *speed*, nor the *chronology* of defining well-differentiated couples of syntactic and semantic classes is relevant for a system to be formal (resp. formally established). As regards the *speed*: it simply does not matter whether the syntactic and semantic profile of a system has been prefixed decades ago or ad hoc specified minutes ago; what counts is that it is possible to draw tentatively on clearly outlined and distinctly interpreted points of reference (e.g. in contexts of arguments). Whether or not this makes sense in the course of a certain procedure is a methodological problem, *not* necessarily a problem of formality. Moreover, as regards chronology: the mode of specification can be *transferred* from an ad hoc approach into a predefined approach in future. Things start becoming more schematic and rigid in this sense, whenever a certain (e.g. scientific) community strictly adheres to a certain system or reading of it, until a better one is proposed. Yet, what changes in the case of internalisation is the degree of familiarity rather than the mode of approaching the system along formal terms, that is, by recursive reference to well-differentiated rules of coupling (whether ad hoc or not).

Finally, I would like to underline that, in consideration of the empirical reference base (established earlier in this thesis; cf. III), it seems to be downright *necessary* to keep the dedicated framework (resp. its modules and variables) open for ad hoc specifications, because only this allows one to illuminate how scientists can actually *learn* (and in this sense obtain epistemically weighty material) from *changes in the way they approach things formally*. This counts for changes in reference after double-checking the analysis of certain systems as well as for changes in the course of time or by the shift of contexts (cf. also 4.3.2.1). What this is meant to say, finally, is that the whole idea of formality employed here depends a great deal on the efficiency of measures that are designed to link up or separate reference systems from each other (e.g. by the use of ostensive signifiers or graphical attractors or distractors, etc.; cf. Fig. 35; cf. 4.2.5.2), but *that* the systems are still open for revision is what makes them appropriate as means in the considered contexts.

After all, what appeared to be a drawback first – namely that ad hoc syntaxes cannot be incorporated into *predefined* settings – leads to an essential pragmatic extension of the framework. The latter would otherwise be too inflexible to illuminate the targeted practices, as it could not integrate the picture-aided *conditions* of establishing formal systems.

4.2.5.5 Differences to other approaches

Let us finally check how the revised view of Perini's account differs from other accounts and contributions in this area. One thing can be clarified in advance: the pragmatic foundations of picture-like means (esp. in terms of reference-dependent traits, test run characterisations and cross-contextual auxiliary systems) have been the centre of focus before (cf. Krämer 2004; Wulf & Zirfas 2005; Seja 2009, to name only a few). However, I claim that the suggested account makes a difference anyway.

This is, first of all, because the account is designed to overcome an awkward separation of research domains: too often it seems as though aesthetics focuses on what makes a picture, while the philosophy of science is concerned with the science-specific exploitation of such ready-made devices. However, a closer look reveals that these two perspectives (which some distinguish as the context of "design" ["*Entstehungszusammenhang*"] and "use" ["*Verwendungszusammenhang*"]; cf. Scholz 2004) are deeply intertwined, which is one reason (among others; cf. 4.1) why I chose Perini as a reference author. For her it is clear that the epistemic weight of pictorial means depends on their nature just as the job-related aptitude of a researcher depends on personal qualities. In this spirit, the suggested framework differs from others already in its strategic focus, insofar as it aims for a synopsis of practically grounded, science-related and still conceptually focused views of pictorial means, which would otherwise not be brought in touch with each other.

As for those approaches that *also* aim for a comprehensive view, one of the most refined suggestions comes from S. Seja (2009). In fact, by the choice of terms and targeted scope, Seja actually appears to come very close to the introduction of ad hoc syntaxes.

In particular, the task of her work is to critically discuss whether and in what respect it makes sense at all to distinguish between *picture-like measures* and *measures that make use of picture-like measures*. As she puts it:

> "The question at stake is, what is meant, anyway, when it is bluntly insisted that one can perform picture-like measures ["*Bildhandlungen vollziehen*"] and use pictures ["*Bilder gebrauchen*]"?"[88]

In the course of her literature review, Seja comes to distinguish between different types of frameworks, with reference to which the initial question is answered differently.

One of those types of frameworks launches the notion of *test run uses of interactively augmented pictorial means* ("*Probehandlungen mit interaktiven Bildern*", Seja 2009, p. 156ff.). This phrase, which is also found in previous works (with the usual discrepancies in the details; cf. Lechtermann & Morsch 2004; Winckler 2004), especially invokes those semiotically approached systems that are implemented on a computer:

> "With this [i.e. with the term "*Probehandlungen*"], a specific kind of pictorial practice shall be denoted, namely one that occurs in virtual realities […] and that involves only virtual objects of operation."[89]

One option accordingly is to analyse the relation between picture-like and picture-based measures with regard to test run modifications of computer-generated systems.[90]

However, before getting lost in this intricate analysis, one may ask how such test run measures ("*Probehandlungen*") differ from the seemingly familiar *test run syntaxes* that were introduced above (cf. 4.2.5.2ff.). In my eyes, the difference is

88 Translation of the following quote: „Deshalb soll die Frage behandelt werden, was überhaupt gemeint ist, wenn häufig wie selbstverständlich davon gesprochen wird, dass man Bildhandlungen vollziehen und Bilder gebrauchen kann" (Seja 2009, p. 10).

89 Translation of the following quote: „Damit soll eine Art von Bildhandeln bezeichnet werden, das in virtuellen Realitäten [...] stattfindet und bei dem ausschließlich virtuelle Gegenstände als Handlungsobjekte verwendet werden" (Seja 2009, p. 20).

90 I will only briefly indicate why I do not agree with Seja's concept of 'interactive pictures' (resp. images): For the most part, because I consider this concept to be contingent, since other entities are also *modifiable*, which is why I do not agree with the idea that there is a sub-type of 'interactive' images. Even paintings in museums can be modified in a way that leaves a visible feedback, only that the respective person will face a penalty for that. Another critical point is the idea of virtual reality. The latter is specified as having an 'own behaviour' ("*eigenes Verhalten*", Seja 2009, p. 21). Yet, in fact, one could conceive of practically everything as having an 'own behaviour' – including organic and ecological systems, etc. For a more elaborated analysis of the concept, cf. Soraker (2010, 2011); for the history of concepts, cf. Richter (2011).

rather obvious in this particular case; Seja is concerned with *epistemic* rather than *semiotic* test run systems:

> "What matters is that those actions undertaken with the aid of interactive images always involve objects, which do not exist in real reality, but only in virtual reality [...]: In the computer-based simulation of a car accident [for example], one does not smash a real car, but uses an image for that – a virtual pictorial object"[91]

Therefore, while there really might be common aspects between what Seja refers to as 'test run measures' ("*Probehandlungen*") and what I identified as 'ad hoc syntaxes' (e.g. in terms of the *reversibility* and so-supported *heuristic function* of the corresponding measures; cf. Winkler 2004, p. 102f.), there is also a major difference. Whereas I meant to examine test run modifications of *semiotic* (esp. syntactic) systems, Seja sheds light on their test run uses as models; accordingly, different functional programs are invoked by each of these two notions. Yet, since those programs are closely interlinked in practice (as has been stressed on several occasions; cf. e.g. 3.3.1), it should be convenient to convert the relation between the two concepts in terms of a *complementary* one; because Seja herself does not dig deeper into the semiotic underpinnings, whereas I myself left the science-specific applications of semiotic systems (e.g. in the context of "*in silico* experiments" Carusi 2008, p. 246) for additional comments in the later parts (cf. 4.3). Still, and for this reason, this is one example for why the notion of ad hoc syntaxes could be appreciated as a contribution to (rather than a mere echo of) recent literature.

This applies similarly to two other accounts, which seem to come close, but ultimately not close enough, to the notion of ad-hoc syntaxes and which thus allow us to distinguish between the conventional aspects and the novelties about this notion.

One of those accounts is by E. Winsberg, who is interested in the "chain of inferences" (Winsberg 1999, p. 275) involved in the design and execution of numerical simulations. The latter are analysed in terms of a sequence of different modelling steps and correlated methodologies, of which at least one entails the idea of 'ad hoc' measures:

---

91 Translation of the following quote: „Worauf es dagegen bei den mit interaktiven Bildern vollzogenen Handlungen ankommt ist die Tatsache, dass man hierbei grundsätzlich mit Gegenständen handelt, die nicht in der aktuellen Wirklichkeit, sondern ausschließlich in einer virtuellen Realität [...] existieren: Man fährt in der Computersimulation des Autounfalls kein realiter existierendes Auto zu Schrott, sondern verwendet das Bild eines Autos, ein virtuelles Bildobjekt" (Seja 2009, p. 157).

> "The hierarchy [of models] includes a mechanical model, a dynamical model, ad hoc models, a computational model, and finally, a model of the phenomena" (Winsberg 1999, p. 276).

Besides this conceptual and terminological overlapping, Winsberg launches a similar intuition concerning the conditions of epistemic weight, insofar as he assumes that "each of these steps in the modelling process is a step on the way toward an inference" (Winsberg 1999, p. 276f.). Each step would thus be worth considering, as it contributes to building up a reference base for claims and arguments. The latter thus build the reference systems when it comes to determining the epistemic weight of entities (cf. 4.2.1).

What is more, Winsberg even states a close relation between ad hoc models and follow-up visualisations, which are regarded as close neighbours in the corresponding workflow.

Yet, in contrast to ad hoc models, the visualisations get important specifically when it comes to uncovering "geometrical patterns of the dynamical processes" (cf. Winsberg 1999, p. 284), that is, when it comes to analysing the visually rendered solutions of ad hoc specified models with regard to significant features of and amongst the acquired data.[92] But even if the visualisations are thus distinguished from the ad hoc phase of modelling, it would still seem that Winsberg is already well-equipped for the introduction of ad hoc syntaxes, since he reveals the integral methodological importance of ad hoc measures as well as of visualisations. He only misses really inspecting the deeper semiotic dimensions of the data analysis and only scratches on the surface of these categories. This is certainly due to the fact that Winsberg aspires to shed light on the *entire* process "with all of its mediating elements" (Winsberg 1999, p. 289), rather than to zoom in on a few of them. By any means, the introduction of ad hoc syntaxes thus proves again to be a new input against the background of another (only familiar) account.

---

92 I will only briefly indicate why the conceptual separation between 'ad hoc models' and models involving 'visualisations' is not convincing: It makes no sense to me to assume that the exploration and interpretation of the model (resp. its solutions) is located late in the workflow. It rather seems to me that there is a strong interrelation between these and earlier stages of modelling. The interrelation surfaces in cross-contextual feedback (in terms of anticipations, back-references and mutual justifications) throughout the process. On that note, I also do not agree with the idea of a "hierarchy" of models (Winsberg 1999, p. 280). Instead, I would prefer to speak of a 'circle' or 'network' of procedures.

The last approach to be mentioned is even farther away from the scope of this thesis, but since it is the source of the notion of *ostensive signifiers*, it deserves more than a footnote here. Besides, it helps to articulate an important disclaimer.

The corresponding source is again Wittgenstein, especially his later work, where he deals (among other issues) with the puzzle of signification, that is, with the question how signs come to refer to something else. In this context, Wittgenstein launches the idea of ostensive signification, which is accordingly considered as an own type of (supposedly auxiliary) referential system:

> "This [i.e. the act of asking something's name], with its correlate ostensive definition is, […] a language-game on its own. […] And there is also a language-game of inventing a name for something, and hence of saying, 'This is…' and then using the new name"[93] (PI § 27).

Hence, ostensive signification (resp. definition) is analysed in terms of a combination of acts of pointing and acts of naming. These acts are (here as well) not just terminological, but also conceptual, because, by *pointing*, one approaches a *type* of correlate, which is an act of conceptualisation, as is the ad hoc approximation of syntactic classes.

Yet, besides those commonalities, there are also cautious claims as to the explanatory potential of ostensive definitions, which need to be consulted, too: Wittgenstein (resp. his critics) argue that ostensive definition does not *solve* the puzzle of signification, since it is *itself* an instance of signification and thus prone to the very same problems:

> "Neither ostensive definitions nor ordinary pictures [which are allegedly similar to the correlate] will stop the regress of interpretations; both are themselves merely signs that admit of various possible interpretations" (Summerfield 1996, p. 110).

This is why reviewing this approach (i.e. the last of the list in the present chapter) forces us again to recall that ostensive signifiers cannot *replace* further conditions in support of the effectiveness and efficiency of symbol systems, such as congruent (or at least close) contexts, education, training, etc. (see also the section on ambiguity below; cf. 4.3.2.1).

---

93 Translation by G. E. M. Anscombe of the following quote: „Dies [das Fragen nach der Benennung von etwas] und sein Korrelat, die hinweisende Erklärung, ist […] ein eigenes Sprachspiel […]. Und es gibt auch ein Sprachspiel: für etwas einen Namen erfinden. Also zu sagen: „Das heißt...", und nun den neuen Namen zu verwenden" (PU § 27).

This is all very well. But does it have an impact on the proposed account? Before moving on, I will explain why *not*: the question about the epistemic weight of pictorial systems was analysed in terms of the question whether they can be part of arguments. Those conditions in turn were analysed in semiotic respects – stressing mainly that systems need to be closed and well-differentiated. Now, arguing that ad hoc syntaxes and ostensive signifiers support such *semiotic conditions* does not mean to say that they *replace* the trial and error methodology. So, there is no *methodological* or *epistemological evaluation* involved yet. Still, this is compatible with the initial task, which was to illuminate *scientific practices*, and practices can never stop infinite regresses other than by some sort of compensatory strategies, which were exactly meant to be modelled here.

> *I am fortunate in that I am able to build on my library of images [...]. For me, every experience builds on the visual memories I carry from prior experience, and in this way my world continues to grow.*
> Grandin (2006), p. 23f.

4.3 A new account of the epistemic surpluses of picture-like means

In order to account for the *epistemic surpluses* of pictorial means, the focus is now shifted to domain-specific performance issues. The question is, to put it bluntly, what do pictorial means have – in terms of technically or economically expounded traits or potential – that other means do not have, especially those that are defined in opposition to pictorial means, such as linguistic means? One might also ask, more elaborately, why the 'logic of saying' could not simply replace the 'logic of showing' in the context of the targeted practices (cf. Richtmeyer 2009, p. 159). This question assumes that there are system-specific limits to the use of symbols which cannot, or only less satisfyingly, be compensated by a change of the conditions of use and which therefore characterise the type and scope of the tool rather than its modes of use and exploitation.

The overall aim of this chapter is to show that the devised framework also allows us to approach problems in the more applied domains, where a refined conceptual framing may contribute to the operationalization of problems and the evaluation of answers. The idea thus is, as in the portrayed scientific projects (cf. esp. 3.1.8), to examine the framework – the model, if you will – with regard to further implications for the studied field. Likewise, the following examination is more of a trigger of *further* research.

4.3.1 Issues with defining epistemic surpluses

Pinpointing the epistemic surpluses of pictorial means is not trivial. One problem is that these means are often related to other (esp. linguistic) means; and these relations (e.g. in terms of transcriptions or translations) are difficult to model in general terms. Bridgman illustrates this quandary (from a still language-biased angle) with special regards to experiments in quantum physics:

> "In the first place, […] we have the problem of translating the evidence of experiment into [an]other language. […] In the second place, we have the problem of understanding the translated evidence of the experiment" (Bridgman 1927, p. 1).

The challenge accordingly is to *isolate* the epistemic surpluses of pictorial means, while in practice everything seems to be fused into a smooth whole of mutual preconditions.

Without fully resolving this problem, one could respond to it in terms of a *synopsis* of the contexts of design and use. In other words, in order to analyse the epistemic surpluses of pictorial means I will again consider both – semiotically classified traits or capacities and functional programs that are epistemically as well as semiotically dimensioned (cf. 2.2.1.2). This means more concretely that I picked out recent performance issues, which are discussed in at least two domains, namely in the area of conceptualising picture-like systems and in the area of optimising them. These issues mainly include the *reduction of ambiguities* (cf. 4.3.2.1) and the *formalisation of pictorial procedures* (cf. 4.3.2.2). They will also include hopes concerning *spatial efficiency* (cf. 4.3.3.1), *early internalisation* (cf. 4.3.3.2) and *interaction* (cf. 4.3.3.3).

Besides the complexity of the task, a second reason for why it is not trivial to determine the epistemic surpluses of picture-like means is indicated by the motto of this chapter: a statement by T. Grandin (2006) on her personal valuing of images. In this connection, it is important to note that any approach towards the epistemic surpluses of pictorial means (whether in science or elsewhere) is doomed to a restricted, perhaps even personalised scope and thus comes off as an open list of correct answers. This is simply because epistemic surpluses are analysed with respect to methodological terms and standards of justification, which are as variant and variable as the actual spectrum of functional programs and user profiles (cf. 2.2.1.3). Therefore, while the last chapter was devoted to the matching of abstract conditions (esp. of picture-likeness and of having epistemic weight), this one assumes the task to draw something general out of something genuinely specific, namely reports and narratives about successes and failures.

In this context, I would like to add another remark: Precisely because practices vary a lot, I will refrain from devoting the coming sections to *merits* and *flaws* of pictorial means. Such evaluations are considered to be part of the inspected context and not of the tool-set of inspection. Ambiguity, for instance, may pose a problem in a treatise, yet not in a poem. Then again, we know that ambiguity has a place in science, and precision in poetry has its own charms. The following synopsis of arguments is thus case-bound in many respects.

4.3.2 Deficiency arguments

4.3.2.1 Some ambiguities about ambiguity issues

When it comes to comparing semiotic systems as to their efficiency as scientific means, troubles with *ambiguity* are usually brought in. Ironically enough, they are themselves in need of specification. This is not only due to the usual divergence of associated concepts, but also due to the complexity of related taxonomies. If it is true, for example, that "[c]alling a picture sad and calling a picture gray are [...] different ways of classifying it" (Goodman 1976, p. 68), and if it is also *prima facie* true that a symbol system is ambiguous as soon as it allows for one or more – up to indefinitely many – alternatives to *all* such ways of classification, it may also be true that there are diverse ambiguities. Coping with ambiguities could thus mean – as with the ambiguity of the *notion of ambiguity* – to choose or rule out a more specific alternative or scope of alternatives of the yet unspecified ways or notions. Another option is to leave the spectrum unspecified (as some suggest for the notion[s] of an image; cf. 2.1.1). Either way, the problem of ambiguity iterates, as there is no general notion of proper specification or generalisation.

Many would thus agree with Goodman (1976) who claims that ambiguities are best avoided in the context of *habitualised* modes of dealing with possibly ambiguous entities (cf. Goodman 1976, p. 71). Accordingly, whether the use of a symbol or symbol system is ambiguous or not depends on whether it varies by habit or due to the resolving of habits, while only the latter amounts to a shift back or forward into states of "sheer" ambiguity (Goodman 1976, p. 70), namely when it comes to a blending, vanishing or emergence of habits of use. Goodman (1976) explains this, using the example of metaphors:

> "As time goes on, the history may fade and the two uses [of a metaphor] tend to achieve equality and independence; the metaphor freezes, or rather evaporates, and the residue is a pair of literal uses – mere ambiguity instead of metaphor" (Goodman 1976, p. 71).

M. Spencer (2012) coins a similar idea, yet based on the notion of moving symbol systems instead of shifting contexts:

> "[T]he thing about inscriptions is that they can travel. Images travel beyond the site of their generation, and along the way there is the possibility of sloughing off accompanying contextual markers, with a variety of legitimacy-bolstering or undermining effects" (Spencer 2012, p. 38; also Lahsen 2005, esp. p. 901ff.).

U. Eco (1976) even introduces a normative perspective when he stresses that ambiguities can be *purposely created* for the good or bad, the beautiful or the ugly:

> "Semiotically speaking ambiguity must be defined as a mode of violating the rules of the code" (Eco 1976, p. 262).

From this assemblage of quotes we learn that ambiguity occurs when there is no (more) convergence between concurrent uses of symbols or compounds of symbols – be it due to a lack or loss of common reference bases or a diversification thereof – up to cases where one struggles to determine whether an entity is an instance of a symbol system *at all*. Yet, as so often, there are windows of influence and exploitation, if you think of functional programs where the accompanying effects of *confusion* are welcomed:

> "[A]mbiguity is a very important device [...] [I]nstead of producing pure disorder, it focuses my attention and urges me to an interpretative effort [...] it incites me toward the discovery of an unexpected flexibility in the language with which I am dealing" (Eco 1976, p. 263).

In view of such appraisals, which are of course divergent, the follow-up tasks now are to clarify which instances of symbol systems exactly suffer or benefit from ambiguity and whether this specifically affects the efficiency of picture-like means.

Let us start with the 'compositional' (resp. 'syntactic') level. Here, one finds ambiguities in terms of graphical marks that could refer to more than just one syntactic class (to stick with the way Perini 2005 and Goodman 1976 would express it).

In fact, such ambiguities are also allowed for by Perini (2005), only she uses a distinct term to address them (resp. the corresponding symbol systems), namely the classifier 'syntactically non-disjoint' (cf. Perini 2005, p. 270f.). This classifier is solely applied to systems, where "the character assignment of at least some marks will be either undecidable or context-dependent" (cf. Perini 2005, p. 270). This is, for instance, the case when a mark – such as O – instantiates at least two syntactic characters – like **0** *and* **O** (cf. Perini 2005, p. 270). Accordingly, the conditions of syntactic non-disjointedness come actually quite close to those of ambiguity, given that ambiguity is also marked by the conditions (resp. effects) of undecidedness and context-dependency. Still, Perini seems unwilling to support the idea of *syntactic ambiguity*, as she opts for a distinct classifier.

Others, however, insist on this as a subtype, not even a mere analogy, of ambiguity (cf. Eco 1976, p. 262ff.). And, in fact, Perini introduces an example in a later paper, which, also in my eyes, brings the notion of 'non-disjointedness' close to the notion of ambiguity (even against her intentions), insofar as the notion of ambiguity is linked up with the idea of *equally approvable alternatives* of identification. Perini claims in the later paper that "[i]t is not always possible to determine whether two micrographs are identical or not" (Perini 2011, p. 920). This, however, can be well explained with regard to ambiguities on *different* levels: on the one hand, each graphically describable difference between the two micrographs could amount to *syntactic ambiguity*, because a differently shaped mark may belong to the same type of syntactic instance or not (as with the **0** and **O**). It could also or additionally amount to semantic ambiguity, since one may relate the differently shaped parts to different referents, respectively (as, e.g., in the case of microscopy-like images where an ad hoc wisely specified syntactic instance might be associated with a foreground part in one context and with a background part in another of the artificially distracted contexts of interpretation). A third option would be to explain the differences between the micrographs with reference to *alterations* of the entire system in the sense that one of the micrographs is taken from a later or more detailed shot than the other one so that, from a methodologically informed perspective, the differences thus identified require the integration of further or modified syntactic and semantic classifiers due to a change of configurations. This might even be admitted as an unique type of ambiguity, namely an ambiguity which is due to an ill-defined back up rationale. Therefore – and even if this appears as only a terminological quibble – I would like to note that Perini's problem with micrographs can be elucidated not only but also in regard to *at least* one more variation, namely the notion of *syntactic ambiguity*.

But before concluding too soon, I shall complement this by further remarks on *semantic ambiguity*. Semantic ambiguity is given, if the syntactic characters of a system can be interpreted differently, that is, if there is more than one referent to which they could be related *as* (resp. *in their function as*) symbols (cf. Perini 2005, p. 271). At the same time there is a certain lack of guidance for the choice between the alternative interpretations; each of them appears to be possible, since for each of them there is a corresponding habit. For Perini (2005), for instance, "[w]ritten and spoken English is semantically ambiguous; propositional logic is not" (Perini 2005, p. 272). This is, because, in ordinary English, there are equivocations, whereas in propositional logic each syntactic character is ideally linked up with exactly one and the same referent in all contexts.

In fact, picture-like systems are also, and often preferably, claimed to be semantically ambiguous in this sense (cf. 2.3.2, 2.3.3). This is even deemed to be the case in scientific contexts, if you consider, for instance, the colour red, which is used to signify all sorts of things (danger, heat, pressure, particles, data points, etc.). Any such system would easily turn ambiguous, if there were no concurrent measures of alternation, separation or ad hoc specification and thus an enforcement of habits to the benefit of ambiguity reduction. The latter are complemented by measures of clarification, tutoring and education, which are often language-based: As Neurath noted: "A certain degree of linguistic education is normally required [...] to tell what is going on"[94] (Neurath 1991, p. 604).

Yet, when it comes to the second question of this chapter, this still does not justify a ranking with respect to the *efficiency* of the specifying (here: linguistic) and the specified (here: pictorial) systems, since the reverse case is also found justified. Consequently, it is argued that pictorial means help specify linguistic systems, if one thinks of the film version of a novel, which is evaluated as to whether it aptly *specifies* the looks of the protagonists or settings. One may also think of photographic evidence, which is more specific than any description of an observed entity, since descriptions rely on relatively abstract descriptors and classifiers, which are devised to fit other entities as well and which are conversely *not* devised to capture all the details that make up a singularity. Hence, neither in terms of *being ambiguous* nor in terms of *aiding ambiguity reduction*, do either pictorial or linguistic systems have a clear advantage.

And so, to try a conclusion on basis of these impressions: the notion of ambiguity does not help pinpoint the epistemic surpluses of picture-like means, because it is a cross-conceptual vagabond; it touches upon many other categories (e.g. non-disjointedness) and it is therefore not restricted to specific semiotic systems.

---

94 Translation of the following quote: „In der Regel ist ein gewisses Maß an wortsprachlicher Erziehung nötig [...], um zu sagen, worum es geht" (Neurath 1991, p. 604).

4.3.2.2 On the (im-)possibility of formalising pictorial systems

Some argue that pictorial means are specifically deficient with regard to the requirements of scientific practice, if you recall the notes from Lagrange (1997), Duhem (1998) and Cartier (1997), to name but three proponents (cf. 2.3.2; 2.3.3).

Lagrange, for instance, completely refrained from using pictorial means, claiming that they are not "subject to a regular and uniform procedure"[95] (Lagrange 1997, pref.). Accordingly, the detailed worries about ambiguities (cf. 4.3.2.1) and other performance issues, sometimes culminate in the notion that pictorial means cannot be integrated into formal settings at all, which is imputed to their very nature:

> "[T]he factors complicating the reading of a picture are many: the system of representation; its many-variabled and many-valued mappings; its lack of fixity and articulation; and, as just noted, the variability of mapped pictorial properties. And it must be emphasized that the knowledge required in reading a picture, the general knowledge of the system and its application to the picture, is implicit and intuitive not articulate and discursive" (Bach 1970, p. 123).

This is but one listing of flaws and obstacles that are typically attributed to the nature of pictorial means rather than to their usages and that are suspected of undermining the efforts of formal treatment. However, even if the corresponding rationales are offered when one tries to track them down, the arguments are not all too clear (cf. also 4.2.5.4). As in the foregoing chapters, the question is thus whether the formalisation of pictorial measures is actually impossible, or simply consumes more resources compared to other semiotic types. Thus, even though the problem of formality has been touched upon before, namely when it came to discussing the *epistemic weight* of pictorial means (cf. 4.2.5.4), it makes sense to reconsider it in the light of the slightly shifted inquiry on *epistemic surpluses*.

This, first of all, requires recalling what has been stressed before, namely that there are different ideas of what a formal system is (cf. 4.2.5.4). Perini (2005), for instance, held that the mode of establishing or approaching a symbol system is *formal*, as long as it can be recursively approved with regard to a dedicated rule-set (cf. 4.2.1.3). This in turn requires that the system be syntactically discrete and articulate as well as bound by distinct semantic correlates, since only under these conditions, so the argument goes, can one tell whether the system complies with

95 Translation by A. Boissonnade and V. N. Vagliente of the following quote: „des opérations algébrique, assujetties à une marche régulière et uniforme“ (J.-L. Lagrange 1853, pref.).

the given rules or patterns of composition and interpretation (cf. 4.2.1.3). This condition becomes clearer (downright trivial) when rephrasing it from a slightly different perspective: If a symbol system is meant to be approached according to certain *rules*, which *restrict* the options of approaching the system, this means that the system must allow for judging whether this succeeded or not, which is certainly easier when the system does not act against a distinct and constrained appeal. However, *whether* the appeal of a system is deemed sufficiently distinct or constrained also depends on the functional background program (whether scientific, aesthetic, or both), which launches the conditions for treating rivalling or wrong approaches towards a system and which guides the employment of auxiliary systems in case certain approaches need to be fostered or prevented. As could be expected, some practices are coined by less rigid error conditions (e.g. some fields of art), while the error conditions multiply in other fields together with the auxiliary strategies of clarification (e.g. in some areas of science).

What is important in any case is that non-formality is *as much sanctioned as it is cured*, which supports the intuition that there are *no inevitable, irreparable* limits to formality. And, indeed, I assume that the alleged limits to formality (e.g. ambiguity, etc.) are exactly those traits to which formality is *dedicated* as a compensatory strategy. For instance, if the syntactic parts of a system overlap or blend into one another (cf. Bach 1970, p. 125), it should be possible to enforce modes of differentiation, in order to render the formerly ill-differentiated system more tractable in this sense. This, however, seems possible with any starting system, independent of the starting type. The limits of formalisation are thus not system- but rather context-specific, because one can always think of a situation where a system may not be treated in accordance with the rules of a certain game.

> „It seems evident from the possible multi-purpose role lines (or patches) of a given type can serve that the system, even assumed fixed, familiar and formalized, cannot select its domain of application" (Bach 1970, p. 122).

For the reason, I argued that the problem of formality is not a principle one (cf. 4.2.5.4), but perhaps rather one that concerns efficiency.

However, *whether* there really is a deficit in efficiency has not yet been sufficiently answered after this brief recapitulation of the previous argument (cf. 4.2.5.4). In fact, even if we assume that pictorial means can be used in accordance with formal terms, there is still some room for exceptions. Therefore, it is possible that they cannot easily be adapted to the contingencies of *specific* formal systems. And, in this sense, there *are* indeed options for arguing that pictorial systems are particularly non-efficient.

Among the more specific aspects of the problem of formalisation, there is the problem of *quotation.* This is quite rich in consequences for the analysis if you recall that for Tarski (who is one of the major reference authors of Perini 2005), it is indispensible that the ingredients of formal symbol systems can be *quoted*; since, only if they can be quoted *in* the terms of the given rules, can one check whether they agree with these rules or not and, thus, whether they are 'formally true' (cf. Tarski 1956; Perini 2005, p. 273ff.; cf. 4.2.1.3). One may extend this to further types of linguistic self-references:

> "When language is used to attribute properties to language or otherwise theorise about it, a linguistic device is needed that 'turns language on itself'. Quotation is one such device. It is our primary meta-linguistic tool" (Cappelen & Lepore 2012).

On that note, one may ask whether it is also possible to put a picture into quotation marks (or something equally functional) and proceed as if it was a quote.

The first thing that is striking about this question is a lack of related associations or intuitions; at least it seems as though they do not spring to mind very easily, which may be due to the fact that there *is* no local practice and related methodology of quoting pictures, or perhaps that they do not operate under the same terminological umbrella.

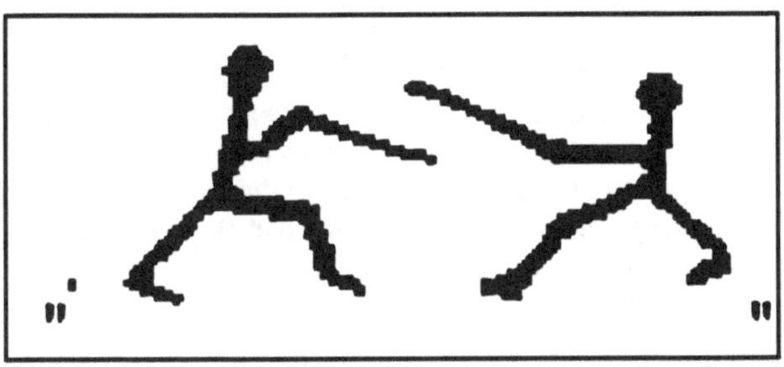

Fig. 37: Extended scope of systematic integration

One finds a rare example in Wittgenstein's notebooks (1961, 29.9.14), if you recall the sketch of the fencing figures (cf. Fig. 37), which is put into quotation marks, and which is referred to as a test run proposition where each syntactic element (resp. each figure that makes a part of the figure) is construed as being a representation in itself. However, even if this example seems to exemplify both a picture-like system and a quotation, it has some flaws from the viewpoint of this chapter. Firstly, it is addressed as if it belongs to the class of linguistic entities, since Wittgenstein calls it a product of writing and further asserts (in accord with his early linguistic bias; cf. 2.3.1), that it "can portray all logical properties of situations in a two-dimensional script" (cf. Wittgenstein 1961, 29.9.14). Therefore, even if one subscribes to the notion that Fig. 37 can be approached either linguistically (e.g. as a set of hieroglyphs; as recommended by Wittgenstein himself; cf. Wittgenstein 1961, 289.9.14) or pictorially (see below), the example might still be a little distractive to a picture-focused inquiry, judged from the way it is embedded in the context of Wittgenstein's remarks.

Secondly, and more problematically, Wittgenstein does not explain the use of quotation marks. Their function is not trivial, though, if one considers the related debate in the philosophy of language (cf. e.g. Cappelen & Lepore 2012; Steinbrenner 2005) and the philosophy of art (cf. Elgin 1983). Yet, whereas in the philosophy of language, the focus is not too surprisingly on linguistic systems, only philosophers of art show any enthusiasm towards the problem of "pictorial quotation" (Elgin 1983, p. 131ff.). And this is also, where the efficiency debate is rooted correspondingly:

> "Some works of art plainly allude to others. But it is less obvious when, if ever, one such work can be said to quote another. Investigation of this question i[s] instructive, for it discloses certain differences in the referential capacities of various symbol systems" (cf. Elgin 1983, p. 131).

At the beginning of her analysis, C. Elgin launches the following intuition on "picturing pictures" (cf. Elgin 1983, p. 132):

> "When a picture is depicted, it is typically shown surrounded by a frame, placed on an easel, or hanging on a wall [...]. So although pictorial systems lack quotation marks, there are pictorial devices that function analogously. Since we can represent any picture by drawing a framed picture that denotes it, both the denotation and the generality requirements on quotation can be satisfied within a representational system" (Elgin 1983, p. 132).

Elgin accordingly assumes two defining conditions of quotation, which, at first glance, can be satisfied by pictorial systems: (1) Quotation implies that the quoted system (or a relevant part of it) is somehow absorbed by and in accordance with the quoting system; and (2) this kind of entanglement is marked as such, while quotation marks are only *one* type of auxiliary tools; there may also be others (such as frames).

This, of course, is but one, probably insufficient concept of quotation. Yet it is intuitive enough to make a point on efficiency that may transcend the scope of this account.

For Elgin, to begin with, "[t]he difficulty in maintaining that pictures quote comes with the containment requirement. [...] A direct verbal quotation contains a syntactic replica of the expression it denotes" (Elgin 1983, p. 132). This condition has not been specified in the above definition, which just states that the quoting system must 'somehow absorb' the quoted system. Yet, this is a point of friction, indeed, since the question is, whether a pictorial system can *entail* another one, while keeping the integrity of both.

This problem – which came up before[96] – can be illuminated by virtue of the devised framework. According to this, if I cannot tell whether a symbol system entails *distinct* parts *before* it absorbs another system of the same indistinct kind, it is also not possible to tell afterwards when faced with the alleged quote, given that the quoting system does not change the type of internal organisation of the quoted system, and vice versa. In more elaborate terms: if there is no alphabet, or any higher order system of the employed syntactic classes (whether ad hoc specified or not) with reference to which a compound of symbols can be successfully broken down into its functional parts, then one cannot guarantee that a quote entails exactly *that* system or *certain* parts if it.

In the face of this quandary, Elgin tries to weaken the conditions of pictorial quotation in the way she compares them with the conditions of "paraphrasing" and "copying" a system (cf. Elgin 1983, p. 133). However, while the analogy to verbal quotation thereby gets weaker and weaker, the problem remains, namely that pictorial quotations do not exactly tell us *what* they rephrase or copy and *that* they do so. Elgin therefore concludes:

> "A picture that contains a copy of another with a frame drawn around it quotes the work it denotes directly or indirectly or not at all, depending on our interpretation of frames, easels, and the like" (Elgin 1983, p. 133).

This conclusion can also be supported with respect to Wittgenstein's sketch (cf. Fig. 37). Wittgenstein propounds that, in order for this sketch to work as a proposition, it must entail at least "everything essential" (Wittgenstein 1961, 29.9.1914). He does not reveal, though, that it may contain *more* than that. But since he assumes that each composing part of the figure obtains a representational function, this opens the door for an indefinite de-composition of the established symbol system (i.e. the arms can be divided into hands, fists, fingers, etc.). What is more, if we thus declare every trait that could alternatively be described in topological, geometrical or graphical terms as a syntactic instance (as done when approaching something *as* a picture; cf. 4.2.2.4, 2.2.2.2), then it becomes difficult to *exclude* possible instances, since even the act of *excluding* or of *turning a blind*

96 This problem was sketched before, when it came to introduce the examples (cf. III intro).

eye to some traits lies within the scope of such 'visual' or 'spatial' description languages. As a result, even a hole in a canvas can be absorbed by the system as a possibly meaningful trait. And if the isolation of functional components is difficult, quoting is difficult too, since the quote can hardly isolate the quoted from the non-quoted (as argued above).

Now, what about less rigid types of representing pictorial means *in* linguistic systems (e.g. for the purpose of analysis)? The reader may remember that I rely on corresponding measures myself, since the challenge was to preserve the integrity of semiotic systems, while transferring them into another (namely this) treatise (cf. III intro.). In this regard, the problem was not so much that the system, which absorbs the other systems, gets semiotically heterogeneous, since one could choose to approach the transferred figures in one or the other way. The problem is rather (as with the 'pictorial quotations') that one needs to communicate *what* has been transferred in terms of traits and interpretations. And this task is indeed prone to efficiency concerns, in terms of making one's choices *better understood.* Related empirical research (e.g. eye-tracking studies) is accordingly devoted to ways of exploring and steering the assimilation of information.

After all, the lesson to be learned is that it is not possible to justify the identity of pictorial means by reference to their traits, because the initial problem iterates with any such act (What traits exactly?). However, since some formal systems require exactly *that* identity, the suspicion remains that systems which build on a discrete catalogue of syntactic classes score better in efficiency when it comes to formalisation. This is because alphabet-like systems are *already* formalised in such a way that they allow for duplications.

4.3.3 Efficiency arguments

The previous chapter commented on some of the most common qualms about the use of pictorial means in science and beyond. Yet, as announced, this is just one side of the coin. Even though the performance-related flaws could not be rejected from out of hand, they were also not completely approved. On the one hand, this was due to the success of compensatory strategies (cf. esp. 4.3.2.2). On the other hand, linguistic means turned out to be prone to the same problems (cf. esp. 4.3.2.1).

I will now add some notes on the current enthusiasm for pictorial means, which is another argument against the listed concerns – not in terms of compensation this time, but in terms of balancing pros and cons.

4.3.3.1 The new master argument: spatial efficiency

The leading question now is why pictorial means are in fact favoured, even if there is a linguistic alternative for conveying the contents?

A widely accepted answer goes as follows: pictorial symbol systems are particularly efficient when it comes to rendering *huge amounts of relationally defined data* (cf. Krempel 2005; Perini 2005a).

The corresponding intuition to capture is that pictorial systems present all contents in one shot, which is usually to say that certain mechanical or behavioural auxiliaries are not needed; for instance, it is not necessary to turn pages or to memorise what has been learned some pages before in order to assimilate what comes some pages later. Thus, even though the notion of *simultaneous* data processing has been rejected in current research, where it is assumed that pictorially rendered data is also processed *sequentially* (cf. Rayner 1998), there is still a strong inkling that pictorial systems are capable of presenting the *whole* of a data base within the instantaneous field of vision. This apparently sounds as if the syntactic body of pictorial systems can be organised in such a way that it allows for a more *concise* semantic annotation. But before I posit a detailed functional explanation for why this is indeed the case, let us see where this surfaces as an advantage in the areas of interest.

There are actually many examples that support the 'spatial efficiency' of picture-like (or at least *rather* picture-like) symbol systems. Consider, for instance, the nomenclature in chemistry. No one is really expected to digest the information that is transmitted by the 'original name' of molecules (also known as the 'compositional' or 'systematic' name), since depending on the size of a molecule this name can fill lines over lines. Therefore, in order to render the classification of molecules more tractable without losing information, *trivial names* are introduced as well as *structural formulas*, which can sometimes even be experienced as three-dimensional (in case it is possible to view them "in a stereoscope, or when 'free-fused'", Perini 2005b, p. 263). The idea is that, thanks to the options of visualisation, it becomes considerably easier to take in the relevant information about the structure of a molecule, such as functional groups and their locations.

Another example is the visualisation of numerical data produced by measurement or simulation (cf. 3.2). Here, an entire cross-disciplinary branch of computer-aided research finds itself downright dependent on the capacity of visual codes:

> "If researchers try to read the data, usually presented as vast numeric matrices, they will take in the information at snail's pace. If the information is rendered graphically, however, they can assimilate it at much faster rate" (Friedhoff & Kiley 1990; cf. 2.3.3).

Clearly, those votes and voices from practice share a certain appreciation of visual means. In this case, the appreciation is justified in psychological terms, since psychology offers categories for the evaluation of visual means in regard to the specific tasks that need to be accomplished by a user in order to complete the functional program of data integration. Concordantly, it is in the light of psychology that visually aided data processing is called *faster*, *easier to learn*, more *comfortable*, perhaps prone to *habitualised interpretations* (making the user *blind* to others), etc. Psychology thus provides classificatory tools that help elucidate the recent keenness as much as the performance-related fears from the perspective of users (resp. user profiles). However, this also means that psychology allows us to capture customised *effects* of visualisation in the first place. From a conceptually orientated angle, the follow-up question would then still be why are there such effects at all? Is there any semiotically supported access to their cause?

This question has been tackled before (cf. Perini 2005a). However, since the proper suggestions rarely enter the philosophical scope of consideration (which might be due to the constant lack of background models, which would permit the assimilation of related arguments; cf. 4.1), they are worth being voiced urgently and in more detail.

From a more granular perspective, the examples turn out to follow the same principles. They are about data sets; and it is important to know, methodologically speaking, how the data points (e.g. the functional groups of the molecules, or measurement values) relate to each other, and what scheme of order they follow in consideration of all relevant relations. In terms of the ordering schemes, there are three types which are specifically deemed suited for visualisation, namely those that are established along *nominal, ordinal or numerical* rule sets (cf. Krempel 2005, p. 17). The latter do not only apply to each of the empirical reference cases tackled in this thesis (cf. III), but also allow us to explain the whole point about the 'spatial efficiency' of visual rendering techniques.

The corresponding efficiency argument first requires, of course, that there is *some* value in marking out such relations in the inspected scientific contexts. Yet, this can easily be shown: consider the virtual cell model (cf. 3.2). The question here is how the signalling particles relate to each other as well as to other parts of the modelled cell when traversing the cell interior. Do they come across other items? How fast do they pass? Do they bustle at some place? Yet, what is thus referred to in the terms of biological events, is actually calculated, thus indeed *numerically* defined, if you recall the unpublished source file underlying the composition of the model, which specifies the relations between the data points in terms of location and speed (cf. Fig. 20). Accordingly, this example is all about the exploration of numerically defined *relations*, even though the exact solutions are not consulted in the end, where it is rather important to pinpoint *tendencies* of system behaviours in compliance with the biological terms of description and scales of interest.

Yet, in order to be suitable for visual renditions, this also requires, secondly, that the listed (i.e. nominal, ordinal and numerical) ordering schemes can be *straightforwardly* set in correspondence with geometrical, topological or graphical schemes of ordering, since this is what efficiency is all about:

> "Visualisations are the more efficient the more unprompted (fast) the conveyed facts can be decoded (read) by the viewer"[97] (cf. Krempel 2005, p. 31).

This can be elucidated as well: take the relation between A and B where A relates to B in the way that A is 'bigger' than B. This relation can be represented by virtue of two marks where one is inscribed as A and the other one as B, and where A is 'bigger' than B in the way that A is different, for instance in size, shape, colour, etc. A difference in *size* may – but need not – come closest to the intuition of applying the attribute 'bigger' and thus yield a faster processing rate. Accordingly, a successful performance only requires that one can use the same descriptor when describing the relation between the two marks as when describing the relation between A and B, namely the descriptor 'bigger' (whatever this signifies exactly). An explication of the conditions of applying the descriptor 'bigger' in both cases should be enough to get this to work. Yet, if these conditions of application are not only settled, but also *aligned*, so that, for instance, the term 'bigger' signifies a difference in size in the case *to be* represented as well as in the case *of* representation, the rendition may become more *efficient* on basis of this convergence, insofar as it renders the use of instructions obsolete for users that know about this convergence.

Depending on further information about A and B – for instance, that they form a part of a cell model – the choice might then be to go for a bigger *square*, or *circle*, or whatever shape is found appropriate to carry the additional information. And given that there is no limit to the addition of further information, this finally indicates that there are indefinitely many options for complicating the conditions and casuistry of visualisation.

97 Translation of the following quote: „Visualisierungen sind umso effektiver, je spontaner (schneller) die dargestellten Sachverhalte von einem Beobachter entschlüsselt (gelesen) werden können" (Krempel 2005, p. 31).

While the defining idea of visualisation might be more palpable now, the question remains why there should be any gap of efficiency between pictorial and linguistic coding systems, because, so far, the revealed gaps concerned different *visual* codes in general.

Further efforts of analysis are therefore needed. In order to hold that linguistic codes *are* more resource-consuming, consider, the following example: *A is left of B and both are above C*. Apparently, I need ***9*** syntactic units (words) and ***26*** atomic characters (letters) to convey that content, which could also be presented by only three syntactic instances:

A B
** C**

This version can even be rendered less resource consuming when one opens up the palette of syntactic characters and chooses smaller ones instead. In addition to this, one could also shorten the distance between the characters:

• •
 •

On that account, the semiotic underpinnings of what I refer to as 'spatial efficiency' can be summed up as follows. The pictorial mode of encoding allows for

(1) comparatively smaller syntactic components, if one admits that every pixel, or whatever is identified as the smallest graphical grain, can be used as one.
(2) comparatively more room for storing higher amounts of syntactic instances. Different factors (including [1]) add to this special capacity: First, it is possible to *arrange* the syntactic units *densely* and, thus, to keep the distances between them short and resource saving. In the extreme, genuinely pictorial case, the syntactic units cannot be differentiated from each other, making the system syntactically non-articulate, while it might still be semantically articulate (cf. e.g. Fig. 13). This becomes extra special if you add the second factor, namely that there are no constraints as to the entanglement of syntactic units, which leaves it up to the user to subsume the same graphical marks to different instances (e.g. a pixel to an atom or to a molecule or to a tumour cell; cf. Fig. 26), depending on which of the neighbour instances are joint together in order to build up more complex syntactic units. This further explains why it is possible to annotate more units in a smaller space. However, it also makes clear that the benefits of spatial efficiency may be lost at the cost of lucidity, and vice versa.

Of course, spatial efficiency requires that one *knows* about a coding scheme (cf. 4.1.4). For instance, one needs to know or be able to justify that 'A is left of

B and both are above C' is the correct reading (see above); otherwise, one may alternatively construe this as 'a triangle-relation between the first letters of the alphabet', etc. Hence, there are often (if not always) more options to describe a range of marks than the one relevant to the accomplishment of a certain (e.g. scientific) task. This, however, is of no harm to the argument of spatial efficiency, since – at least in the inspected scientific cases – there are *conditions of relevance*, which help to discern and justify descriptions as being adequate. These conditions are given by the model to be rendered, which works like a cloze, insofar as it provides some sort of identikit profile of the information of interest. For instance, if one expects a certain particle to move around like the biological reference model, it is possible to assimilate the visualised solutions of a numerical model in that *focused* way.

Based on the semiotic principles (i.e. small-sized syntactic classes and a large palette of concise [up to dense] modes of composition) and the context-related principle that posits essential traits of scientific modelling (i.e. relation-focused analyses), this can now be canalised into a functional explanation of a veritable epistemic surplus. Consider one of the first explanatory attempts in this connection:

> "The reader of a book follows a fixed marching line. Only someone who is well-exercised could get a quick overview over the content of an article or book [...]. [By contrast] in a two-dimensional image, it is possible to "see" several interconnected arrays of objects concurrently, while the one-dimensional text requires that the reader memorises what he has read and that he establishes connection between the single statements he picks up"[98] (cf. Neurath 1991, p. 602).

Perini follows up on this, yet in accord with the terms of her own framework:

> "Because pictorial systems are dense, and because they use visible features to represent properties, they have the capacity to represent a dense set of very complex properties" (Perini 2010, p. 144).

Hence, to draw the comparison again: In order to encode spatially dimensioned relations by virtue of a *linguistic* symbol system, all related components

98 Translation of the following quote: „Der Leser eines Buches hat eine fixierte Marschlinie. Nur jemand mit viel Übung kann sofort einen Gesamtüberblick über den Inhalt eines Artikels oder Buches bekommen. [...] In einem zweidimensionalen Bild können mehrere Reihen von Gegenständen, die miteinander verbunden sind, gleichzeitig „gesehen" werden, während der eindimensionale Text verlangt, dass der Leser im Gedächtnis behält, was er gelesen hat, und die Verbindungen zwischen den verstreuten Einzelaussagen selbst herstellen" (Neurath 1991, p. 602).

have to be named, one after the other, and related to each other, using separate propositions to signify each relation. Each of the syntactic *relata* thus needs to occur (more) often as duplicates and there must even be symbols for stating the relations themselves. By contrast, in a pictorial rendition every syntactic instance needs to occur *only once* in order that it can be put in *different, concise* relations to *all* other instances. And since this eventually comes to light in direct comparison to linguistic means, it may truly be appreciated as an epistemic *surplus* or a compound of surpluses (as to the conditions set up before; cf. 2.2.1.3). Let us now see in the following whether the list of benefits can be further extended or whether other benefits can also be illuminated against the background of this one (cf. esp. 4.3.3.2).

4.3.3.2 Internalised codes and the haunting intuition about similarity

When it comes to pinpointing possible surpluses of pictorial means, spatial efficiency is usually not amongst the first associations, although it can be derived by considering their defining traits. People rather tend to think of their aid in *fostering* and *documenting observations*. However, it made sense to start off with spatial efficiency, since I claim that the following intuitions can also be smoothly elucidated against this background.

The processing of observations is rarely examined as a separate functional program in the philosophy of science. More typically, it is conceived of as an interim module at some point between the acquisition and evaluation of data. However, part of working with data is documenting it for later inspection or memory. This may be comparatively easy when it comes to plotting numerically specified data (e.g. by means of coordinate systems), although corresponding techniques required notable steps of abstraction in history (cf. Vogelgsang 2009); but it becomes intricate as soon as it concerns the macroscopic profile of research objects in geology, medical engineering, etc. Here, researchers also need to work with large amounts of data – yet with the additional challenge that this data is not, or not as easily, ordered according to a nominal, ordinal or numerical scheme, if you think of the morphology of rocks, for instance. Concordantly, these branches perfected the craft of *depiction* (cf. Kemp 1990, 2000, Habbe 1996; cf. 3.3.3). Here as well pictorial means became convenient to use in any respect. But what is it exactly that makes them score better than other semiotic types in this context? It must have to do with the intriguing notion that – to the benefit of observation and documentation – visual representations are particularly 'close' to the target system in regard to the variables they employ and the way they employ them. Amongst the *graphical* variables, there are the ones listed by B. Bertin (cf. Fig. 38):

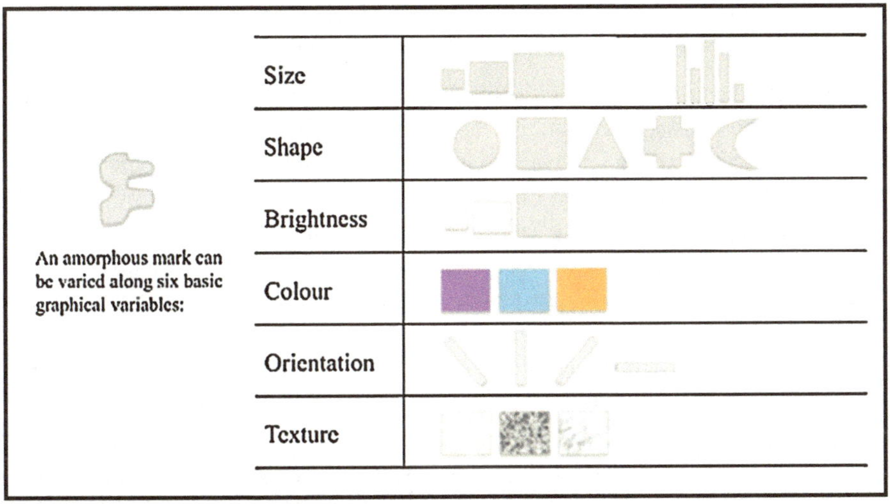

Fig. 38: B. Bertin's graphical variables (cf. Bertin 1974, p. 49ff.)

The striking aspects about these variables can be condensed to a list of two: first of all, their recognition is internalised early in life, since, in contrast to the syntactic components of graphically extended *linguistic* symbol systems (i.e. words and letters), these variables are rudimentary in compositional respects and also relevant to rudimentary behavioural systems, such as navigation; they help humans to find their bearings, to estimate distances and proportions, or to tell between good and bad looking food. The resulting impression is that systems that build on these variables are intuitively more tractable (cf. Krempel 2005, p. 15), which is to say that, once annotated, it is easy to invoke and train the proper reaction (e.g. reading red as signifying danger). And precisely because the recognition of these traits is internalised early, it is also widespread in all areas of society as well as most regions worldwide. The systems that transcend the borders of linguistic literacy can be counted as a second striking aspect, which is why such upshots of scientific practice are more likely to wander through the public media as opposed to other testimonies.

The aspect of early internalisation and the specific compositional traits of pictorial systems (esp. the aspect of spatial efficiency; cf. 4.3.3.1) taken together provide a key to the laudable benefits of pictorial systems in the context of recording observations. The related argument is that visual systems offer *as many* options of the *same type* of composition as the singularities of daily perceptual (esp. optical) experience do.

> „The subject of pictorial representation is the visual form. The visual sense offers the greatest freedom of representation, not only due to the richness of data to choose from, but also due to the amount of variables […]"[99] (Jonas 1961, p. 165).

Due to both the compositional resources and the human conditions of processing them on a daily basis, there is thus reason for the suspicion that the experience of visual representations can *converge* with perceptual (esp. optical) experiences in the sense that the same optical inputs can be encoded in the same way with only minor losses. In effect, the experiences may appear intriguingly *similar*, which is why photography, microscopy and other visual devices have gained a good reputation in science, where "the significance of depiction [is that of] an aid to extraction" (cf. Perini 2012a, p. 151).

On this basis, some even go so far as to blend the epistemic weight of perceptual (optical) experiences and the epistemic weight of visual symbol systems:

> "Maybe the reliance on pictures is symptomatic of a shared cognitive limitation we humans have or maybe we should cast this reliance in a more positive light and think of it as part of the standard machinery for […] human understanding" (Downes 2012, p. 115f.).

This risky analogy is still helpful, though, for it helps to reveal an as yet unmentioned flaw, which I have skipped until now, because I consider it a problem not of the nature of images, but of their unexercised, over-trustful use:

> "[I]mages are seductive because we are such visual beings. All sighted humans interact with our world via vision more so than any of the other senses. When an image is presented to us, we automatically think that it tells us what something looks like" (Downes 2012, p. 124).

G. Bachelard bemoaned the dangers of an unwary processing of visual(ised) inputs against the background of scientific virtues. His position is cited correspondingly:

> "If science is to progress, it is necessary that scientists disrupt imaginative faculties, that they find ways to kick the feet out from under intuitive ways of thinking that have been cultivated across the many arenas of daily life. It is essential for modern science, in Bachelard's view, to overcome what he calls 'epistemological obstacles'" (Spencer 2012, p. 36).

---

99 Translation of the following quote: „Gegenstand bildlicher Darstellung ist visuelle Form. Der Gesichtssinn gewahrt die größte Freiheit der Repräsentation, nicht nur durch den Reichtum der Daten, die zur Auswahl stehlen, sondern auch durch die Zahl der Variablen […]" (Jonas 1961, p. 165).

In this sense, the opponents of the scientific exploitation of pictorial means have a good point, indeed (cf. 2.3.3; 4.3.2.3). The problem is that they believe that this is all about the *lack* of formal constraints, while it is actually a problem of not letting oneself be distracted from formal constraints by a flood of associations and inferences.

However, as to be expected from the diversity of functional programs, 'wrong' associations are not always seen as a problem. Even in science, the associations are played with, if you recall Waddington's epigenetic landscape (cf. Fig. 33). The latter illustrates intra-cellular decision-making only if one succeeds in linking it up with the uncontrolled, mono-directional behaviour of a rolling marble, which has nothing to do with other cell models.

But before getting lost in problems of casuistry, the common point of friction seems still to be – even more in the face of the diversity of examples – this notion of being 'closer' to visual experiences. On the one hand, this is a candidate for a surplus; on the other hand, the accompanying notion of similarity has apparently stirred heated arguments.

Goodman, for instance, furthered the debate in the way that he showed that the concept of similarity does not work as a defining condition or trait of pictorial representation (cf. Goodman 1976, p. 3ff.). This is, in a nutshell, because the options of *comparison* are indefinite and so are the options of finding similarities: a *fruit* and a *legume*, for instance, are different in respect of botanic taxonomies, but they are similar in the way that they can be eaten, or in the way that they contain water or in the way that they are part of our world. In the end, everything can be similar to anything, given that one can always define a set, which includes both of the compared entities. And, on that note, the question indeed is: what causes the haunting effect of similarity in the case of pictorial means?

I have briefly indicated a direction of answering this question here; but before I move on to the next chapter, I would like to propose a recapitulation in more abstract terms.

Most important perhaps, I suggest defining the advantage of similarity as an advantage of *efficiency* rather than *effectiveness*. Since if similarity is defined in terms of category matching then similarity stretches over multi-modal perceptual input, meaning it can also apply to acoustic, tactile, etc., features of the entities compared. Consequently, one may detect similarities anywhere without even being entirely conscious about them (not to mention putting them into words). Therefore, instead of hypostasising similarities – that is, thinking of them as being specific traits or surpluses of pictorial means – I suggest that the latter are particularly good in encoding observations, thanks to the capacities noted above.

The one revealed in this section is that pictorial means store data of a type that is internalised very early in life as part of the optically aided navigation system; this is why information encoded in this way can be assimilated into a larger memory of experiences and exercise. And, secondly, the pictorial type can store this information very efficiently (cf. 4.3.3.1), meaning that these means perform comparatively well when it comes to keeping up with the variety and density of everyday optical experience. Thus, while the epistemological notion of 'being closer to reality' is rejected, the semiotic surplus of 'being closer to a certain type of presenting or assuming content' is preserved on that score.

4.3.3.3 Refining the notion of interaction

There is another aspect – especially with the *computer-aided* use of pictorial means – which is also found addressed in terms of a surplus, namely the fact that these means allow for *interaction*. In visualisation practice, this is articulated, for example, as follows:

> "[A]s the simulation usually covers several hundred frames, we allow the user to step through them by keystrokes. The position is interpolated for a smooth motion of the molecules between two frames. [...] Up to three clip planes can be enabled to cut away uninteresting or occluding parts of the data set. Additionally, the clip planes can be used to prepare slices by aligning them in parallel [...] Camera parameters like position, orientation, field of view, and the aperture are to be set by the user [...]. The colours can be adjusted by the user so that they match, e.g., images obtained by wet lab experiments" (cf. Falk et al. 2009, p. 174).

The authors thus posit a range of operations that can be performed either with the pictorial symbol system itself, or with the results of studying it, in any case *thanks* to the specific type or type-specific equipment of the particular symbol system. In that case, however, the underlying idea of *interaction* calls for a further analysis, since there seem to be several notions fused under this common terminological umbrella.

The notion of interaction, to start from a generic angle, seems derived from the interaction between persons, where each is an actor and where each action has the logical status of a *reaction* to the presence or previous moves of the other person. Along the lines of such a *request-response model*, the above quotes support at least two semiotically dimensioned analogies underlying the talk of human-system *interaction.*

The interaction between a user and a symbol system could, first, be analysed in terms of an *intervention.* Say, the user intervenes into the composition of a symbol system and the system – in some sense of a reaction – adopts the intervention in terms of an adjustment of its syntactic body. The user in turn can react to how this comes off, etc. On that note, one could even further distinguish between a weak and a strong version of this type of interaction (while only the strong one actually operates under this term): *weak* interaction in this sense amounts to the refinement of interim graphical renditions (e.g. for purposes of publication; cf. close-up analysis of Fig. 13). This is a daily routine in the areas under study given that it is not too different from the refinement of typefaces under criteria of *legibility*. By contrast, the *strong* idea of interaction seems a little more system-specific. It concerns, for instance, the use of the virtual cell model introduced earlier in this thesis (cf. Falk et al. 2009; cf. esp. 3.2.12). Here, the user is equipped with a whole palette of options of intervention. These options comprise a broad set of graphical alterations and further measures of (re-)defining the relations amongst these variables, including the choice of perspective-parameters. By dint of the efforts spent on making such modes of change available in terms of *options of change for any user*, this apparently extends the idea of mere refinement of what is approached in terms of the syntactic body of a figure.

Accordingly, the first notion of interaction amounts in both cases (i.e. the weak and the strong one) to a more or less technically aided interplay between the changing of symbolic (esp. syntactic) corpora and the processing of optical feedback. The optical feedback in turn is manifested in accord with the rules of the semiotic system, meaning that the system 'reacts' to changes in accord with its own (resp. with the ascribed) mode of 'parlance'. This may appear as a trivial add-on to the foregoing analysis; however, there is more about this finding in regard to the notorious dichotomy between linguistic and pictorial means. Given that pictorial means build on functional parts, which can alternatively be described by virtue of geometrical, topological or graphical classifiers, and given that these functional parts are densely arranged (as is typical of this semiotic type; cf. 4.2.2.4), they have the capacity to absorb *any* such dimensioned intervention that enters the frame of their particular architecture – from an accidental scratch on a paper all the way to a bullet shot into a canvas. Hence, on account of the defining conditions of picture-likeness, *any* newly involved trait easily turns into a syntactic unit and thus offers itself to be coupled to a semantic unit. By contrast – and this is where another, not instantly epistemic, but at least semiotic surplus looms on the horizon – a textual corpus can only absorb graphical changes under peculiar conditions; for instance, when changing an **F** into an **E** or when giving specific colours to a (e.g. hyper-)text or parts thereof. Not forgetting that this *also* requires that the entire syntactic context (e.g. a word or sentence) then makes still sense, which in turn restricts the scope of meaningful interventions. Ultimately, it thus turns out (resp. shall be stated here) that the intervention into textual systems – by virtue of rudimentary, navigation-like 'moves' – is comparatively rich in preconditions.

The second notion of interaction, to continue with, builds on the previously analysed idea of syntactic interventions, but comes out more specifically, as it profiles the interventions with special regards to the functional program of modelling from which it adopts further conditions. Consider, for instance, E. H. Hutten (1956) in this context:

> "The pictures we draw for a simple experiment – say, a collision between two particles – are not merely used for supporting our limited imagination. The geometry of the situation, the lines representing the forces, the angles between the lines marking a direction, etc., allow us to use a vector calculus in order to solve the problem" (Hutten 1956, p. 85).

This quote is more of an illustration than a veritable instance of the talk of interaction, but it helps pinpoint further defining conditions implicitly conveyed by this talk (e.g. by the initial quote of this section). Hutten first of all invokes the previously analysed notion of intervention insofar as he holds that data sets (produced, e.g., in the course of an experiment or a simulation) are presented and modified in a semiotically classifiable way. He then adds a functional background where the individual measures are systematically fused and subsumed under the shared idea of an intermediate step of research. This intermediate step of research is often characterised with an enabling effect (esp. concerning the acquisition of hypotheses) and, in this connection, also with a mediating function (esp. concerning the post- and pre-processing of data):

> "For a scientist, the figuration of his model may bridge between the expectations triggered by the calculus and the support gained by experiments"[100] (Boehm 2007, p. 37).

In the end, the analysis of the second notion of interaction thus comes off as an appraisal of the *heuristic* function of pictorial systems in the context of data exploration (cf. also Kötter 2007, p. 178). Still, one may finally uphold that the specific merits of pictorial means in fact amount to those of spatial efficiency (cf. 4.3.3.1) and early internalisation (cf. 4.3.3.2), as the latter help to explain why interventions can be efficiently absorbed by the type of system and further processed by the 'type' of *human* user. Thus, *interaction* does not extend but *systematically enhances* the surpluses of pictorial systems.

---

100 Translation of the following quote: „Für einen Wissenschaftler mag die Figuration seines Modells die Lücke schließen zwischen dem, was ein Kalkül erwarten lässt und dem, was die Erprobung bestätigen kann“ (Boehm 2007, p. 37).

V Outlook

The aim of this work was to offer an extended framework that accounts for the epistemic weight and epistemic surpluses of genuinely picture-like means, since – in stark contrast to their prevalence in scientific practice – these means have long been left aside in the philosophy of science. Among the rare explicit reasons for this neglect, the most salient one is that the here adopted *definiens* of picture-like means – syntactic density – directly rivals the syntactic requirements of arguments, namely with being composed of well-differentiated atomic parts which can be combined and recursively enhanced according to distinct rules. But this conflict can be resolved which is what I intended to show by taking into account *ad hoc syntaxes*. The latter allow for a redefinition of picture-like means in a way that reveals them as being formally tractable while keeping their integrity as picture-like means. Assuming that this has been successful, it still remains possible to continue the efforts of framework building at different points. Thus, I would like to close with some hints to further options for research.

First of all, the defining conditions of each conceptual variable could certainly be analysed in terms of their own defining conditions, etc., which might foster the integration and acceptance of this issue in the more general domains of philosophy, insofar as this would build a bridge to more fundamental issues (e.g. perception). Secondly, it is likewise possible to extend the empirical reference base – either in terms of the documentation of further research practices or in terms of the involvement of results from experiments (esp. eye tracking studies). This would put the framework to the test as to its scope of empirical significance. And, thirdly, there is the option of refining the exegeses, since, as can be expected, the adopted reference accounts (esp. Goodman 1976 and Cummins 1975, which lead to the appreciation of Perini 2002ff.) are controversial – albeit defensible and very profitable, as I maintain without concessions.

In terms of the integration of additional variables, some suggestions stand to reason: it would be worth adding an analysis of the conditions of *perception*. This analysis would prepare the framework to be discussed and enhanced against cognitive views of data assimilation and concept building (cf. e.g. Pompe 2011). This could particularly promote the search for a more elaborate notion of *ad hoc syntaxes*. Moreover, there is the option of getting involved in sociological studies, which investigate the social conditions and local peculiarities of picture-aided research and conditions of efficiency. Last but not least, there is room and reason for an extension of historical embedding. I started the literature inquiry with works from the early 20th century (with only a few exceptions), since I did not want to get too far away from current intuitions on what science is. Nevertheless, it should be possible to continue the inquiry in the course of extended historical studies.

Irrespective of the various options for continuation, of which there are certainly many, I would like to stress again that the suggested framework is meant to provide sufficient material to meet the initial tasks of this thesis, namely the systematic integration of the epistemic weight and surpluses of pictorial means. In this context, several aspects should be worth mentioning again, because, facing the fact that the initial problems have been tackled before, there are some aspects that I wanted to improve *specifically*.

For instance, I gained the impression that it is difficult to integrate pictorial means into a treatise about their methodological status and, at the same time, not let them appear as obstacles in the flow of the text, ending up with some kind of performative contradiction. The strategy I therefore chose was to involve *close-up analyses* as well as accompanying reflections on the quandary of representing representations. Furthermore, and especially since this work was done in an interdisciplinary context, my aim was also explicitly to motivate the necessity of a synopsis of philosophical, historical and empirical accounts, particularly when it comes to analysing functional programs in science, which are needed to model the potential scope and requirements of pictorial practices. In this context, it was also my aim to stress the entanglement between methodological, syntactic and semantic modes of establishing or approaching symbol systems and to argue on that score that not every computer-aided system automatically fits the criteria of language-likeness. On the contrary, this led the previous discussion astray and made the proponents blind to genuinely pictorial *yet still* discursively integral measures such as the introduction of ad hoc syntaxes and the accompanying techniques of formalisation. Apart from this extension of the previous reference models, my aim was to test the devised framework with regard to its implications on domain-specific performance issues. This helped to further flesh out the previous analysis, since the efficiency debate profits from and gives feedback on the semiotic reflections. Having faced all these aims and expectations, I claim to have offered an admissible solution.

Summary (English)

In the first part of this thesis, an account of the main problem definitions, motivators and methodological underpinnings was presented. I particularly stressed in this context that this work is in line with a broader movement in the more recent philosophy of science, insofar as it shares with other works a certain starting point, which it acts against: and that is the purely linguistic view of *representation*, and here especially also of *argumentation*. This work is dedicated to the counter-intuition, which holds that non-linguistic – and especially *pictorial* – means *can* be taken into account when it comes to illuminating what science is in terms of how representations and argumentations work in functional regards and under -science-specific demands on the performance of these means. Conversely, a lack of integrative efforts concerning the role of pictorial means in contexts of representation and argumentation was taken as a shortcoming of previous approaches.

The second part of this thesis presented two close-up analyses of the targeted scientific practices, some of which have not been documented in the philosophy of science so far – such as the use of microscopy-like images. The main challenge was to operationalize the questions on the epistemic weight (i.e. the functional role) and system-specific surpluses (i.e. type-specific contributions) of pictorial means in such a way that it becomes possible to decide on a case basis whether something is *picture-like*, *language-like*, *epistemically weighty* or *particularly efficient*, etc. At the same time, these conditions should, of course, allow for the intuitions gained from the inspection of the reference examples. The most important intuitions gathered in this context were that picture-like means can be a non-trivial part of genuinely 'scientific' or 'discursive' measures, such as representation and argumentation, and that they can be indispensible due to their distinct capacities.

In search of previous models of the state and status of picture-like means in the targeted contexts, the literature inquiry revealed notorious deficits. Two of them suffice to justify an analysis of the extent of this thesis: First of all, it turned out that a satisfying answer to the problem of the *epistemic weight* of pictorial means requires the clarification of *four* interrelated *conceptual variables*, namely of the concept of *picture-likeness* as opposed to the concept of *language-likeness*, and of the functional conditions of *picture-based* measures as opposed to the functional conditions *language-based* measures. In this context, the literature survey revealed that most of the recent contributions focus on only *some* of these variables, and usually not on the measures of interest, which makes the accounts feeble and too limited in reach with regard to the unclarified *dichotomies* between picture- and language-likeness on the one hand, and between epistemically weighty and non-weighty items on the other hand.

The second deficit of the reviewed accounts comes with the distinction between the context of design and use. Due to hypostatisation – that is, due to taking this distinction as a trait of practice rather than as an upshot or tool of the analysis – the distinction has led to an insupportable neglect of the inter-relations between compositional, representational and methodological levels of reference, which are deeply intertwined in practice. The hypostatisation of these distinctions accordingly comes off as a veritable systematic deficit, considering, for instance, the resulting conviction that computer-generated systems *principally* fit the conditions of language-likeness. To avoid such semi-empirical dubieties, a synopsis of the context of design and use is inevitable.

With these aspirations and demands in mind, L. Perini's (2002ff.) work was picked as a point of reference and departure. The salient aspect about her work (which was much inspired by Goodman 1976 and Cummins 1956) is that she integrates *all four* conceptual variables that are needed to tackle the initial problems. Besides, she takes into account examples from the broader domains of micro- and systems-biology, which helped the alignment of the empirically inspired intuitions on the issues. At the same time, this hinders neither the transfer of her nor of my revised account to *other* domains of scientific visualisation thanks to the general *semiotic* ingredients of the frameworks and thanks to the *variety* of picture-aided practices within the inspected cases and fields themselves.

In spite of these virtues of Perini's ambitious agenda and approach, her work in this field nevertheless also revealed crucial deficits: the most intricate problem with her account is that it is not capable of integrating *genuinely pictorial* means into the scope of conceptual reach. Accordingly, on the one hand, she admits these means play a significant role in the scientific contexts of interest. On the other hand, the framework fails to integrate them systematically. This is because the major bridging principle employed in her account – *translatability* – does not allow for the compositional trait of *syntactic density* (Goodman 1976), leaving the account ultimately 'blind' (resp. 'silent') to the epistemic weight of the so-defined *genuinely pictorial* means.

In the subsequent sections, this has been assumed as a task. In this context, it turned out that the systematic integration of genuinely pictorial means is attainable without being in need of any stipulations or disclaimers concerning the expectations on a solid account of the epistemic weight and surpluses of picture-like means. In fact, this is intended to be the central position launched and defended by this work in reply to these two problems.

In order to further substantiate the related theses on the epistemic weight and surpluses of pictorial means, the introduction of *further* conceptual analyses, bridging principles and empirically inspired conditions was necessary. Amongst these additional conceptual ingredients, there are the concepts of *ad hoc syntaxes* and *ostentatious signification*, which ultimately allow us to elucidate how genuinely pictorial systems can indeed be functionally involved in the targeted contexts in a non-trivial way while maintaining their integrity as *pictorial* means and while maintaining the elaborate functional analysis of these discursive contexts – that is, of representation and especially argumentation. At the same time, the conceptual ingredients which are introduced help to shed light on conditions of formal practices, which have been neglected by the accounts under consideration, but which are necessary in order to make sense of the targeted practices.

On the basis of the devised framework, the second part of the analysis shifted the focus to rather domain-specific performance issues. The idea was to continue the analysis with regard to problems of application that are presently discussed at the interface(s) between philosophy and scientific visualisation. In this context, the framework revealed helpful implications; so that some of the reviewed arguments could be either rejected (esp. the deficiency arguments on *ambiguity*) or refined (esp. the efficiency arguments that build upon accounts of *similarity*, *interaction*, etc.). Veritable surpluses were found in terms of *spatial efficiency* and *early internalisation*. In order to corroborate these revisions of previous technical and economical arguments from the suggested conceptual viewpoint – and with this the claim that they *better* illuminate the targeted practices – the thesis is equipped with an extensive empirical reference base. New inputs could explored in this context thanks to close cooperation with the local scientific and engineering staff.

Summary (German)

Diese Arbeit ist wie andere wissenschaftstheoretische Forschungsarbeiten der jüngeren Zeit aus der Überzeugung heraus entstanden, dass sich diejenigen Verfahren, von denen – ungeachtet aller Kontingenzen der wissenschaftlichen Praxis – angenommen wird, dass sie zentrale funktionale Module wissenschaftlicher Forschung und Wissensbildung darstellen, nämlich die *Darstellung* beziehungsweise die *Repräsentation* selbst und die *Argumentation*, nicht auf sprachliche Instanzen reduzieren lassen. Kurz gesagt: Ein rein sprachbasierter Begriff von Wissenschaft scheint seinem Gegenstand nicht gerecht zu werden. Diese Arbeit ist daher der Gegenintuition gewidmet: Es wird gezeigt, dass non-verbale und unter diesen vor allem *bildliche* Mittel zu berücksichtigen sind, und tatsächlich auch berücksichtigt werden *können*, wenn es darum geht, die genannten Verfahren so zu modellieren, dass sich Aussagen über die funktionale Integrierbarkeit unterschiedlicher Symbolsysteme (das heißt über deren epistemischen Status) und über systemspezifische Leistungen (das epistemische Surplus) treffen lassen.

Die große Herausforderung dieser Forschungsarbeit war es, die beiden Fragen nach dem epistemischen Status (*epistemic weight*) und den systemspezifischen Leistungen (*epistemic surpluses*) bildlicher Mittel zu operationalisieren. Es ging also darum, die in den beiden Fragen genannten begrifflichen Variablen in ihrem Verhältnis zueinander so zu bestimmen, dass sich konkrete Bedingungen angeben lassen, unter denen sie gelten, das heißt unter denen etwas *bildlich*, *systematisch relevant* und *besonders leistungsfähig* ist. Gleichzeitig sollten diese Bedingungen auch so bestimmt sein, dass sie Beispiele aus der Praxis beziehungsweise die aus der Erfahrung gewonnenen Intuitionen absorbierbar und begreiflich machen können. Auf einer solchen Grundlage, so war die Idee, könnte, gemünzt auf den jeweiligen Bezugskontext, für oder gegen den epistemischen Status bildlicher Symbolsysteme sowie – gegebenenfalls unter Berücksichtigung weiterer Bedingungen – für oder gegen systemspezifische Leistungen argumentiert werden. Folglich wäre auch zu erwarten, dass sich auf dieser Grundlage die *Defizite* bisheriger Argumente hinsichtlich ihrer Schlüssigkeit oder ihrer empirischen Reichweite bestimmen und als Anschlussprobleme formulieren lassen.

Diese Ziele und Erwartungen wurden auf der Grundlage einer Literaturrecherche und -kritik formuliert, die sowohl einen methodologischen Fokus hatte, das heißt Varianten der philosophischen Erschließung wissenschaftlicher Praxen wenigstens kurz diskutiert, als auch einen begrifflichen Fokus, also bisherige Argumente zum Status von Bildern in den Wissenschaften bespricht. Bei der Literaturrecherche haben sich Defizite bei der Integration *genuin bildlicher Mittel* durch bisherige wissenschaftstheoretische Ansätze erkennen lassen und zur Motivation dieser Arbeit beigetragen.

Unter diesen Defiziten sind mindestens zwei zu nennen, die eine Untersuchung vom Umfang dieser Forschungsarbeit begründen können. Die Literaturdurchsicht hat erstens ergeben, dass Argumente zum epistemischen Status bildlicher Mittel auf mindestens *vier begriffliche Variablen* beziehungsweise *Analysen* Bezug nehmen müssen. Dazu gehören der Begriff der *Bildlichkeit*, der im Verhältnis zum Begriff der *Sprachlichkeit* zu bestimmen ist. Und dazu gehören die funktionalen, häufig definierenden Bedingungen *bildbasierter Verfahren*, die im Vergleich zu den funktionalen Bedingungen *sprachbasierter Verfahren* zu analysieren sind, insbesondere mit Blick auf die Verfahren der Repräsentation und der Argumentation. In diesem Kontext wurde festgestellt, dass die bisherigen Beiträge mit wenigen Ausnahmen mindestens eine dieser Variablen und damit auch deren Verhältnis zueinander vernachlässigen. Infolgedessen werden Aussagen über den epistemischen Status bildlicher Symbolsysteme anfällig für Begründungsprobleme, und zwar zum einen bezüglich der Bild-Sprach-Dichotomie und zum anderen bezüglich der Dichotomie zwischen integrierbaren und nicht integrierbaren Komponenten. Wenn aber der Anspruch besteht, die Ausgangsfragen problemadäquat zu diskutieren, das heißt mit Blick auf die *problematisierten Dichotomien*, dann fällt dieses Defizit ins Gewicht.

Das zweite Defizit besteht in der Hypostasierung der Unterscheidung zwischen dem Entstehungs- und dem Verwendungskontext bildlicher Systeme. Diese Unterscheidung mag zwar zwischen verschiedenen Arten der Bezugnahme auf bildliche Symbolsysteme richtig zu trennen. Aber wenn man dies nicht als Ertrag der Analyse, sondern als Praxismodell missversteht, dann verstellt diese Unterscheidung den Blick für den Zusammenhang zwischen kompositorischen, repräsentationalen und methodologischen Ebenen der Bezugnahme, die im Umgang mit Bildern immer eine Rolle spielen, das heißt praktisch nicht zu trennen, sondern systematisch aufeinander bezogen sind. Als Defizit äußert sich die Hypostasierung vor allem in Form von zweifelhaften, semi-empirischen Annahmen wie der, dass computergenerierte Symbolsysteme *prinzipiell* sprachlicher Natur sind.

Angesichts dieser Defizite wurde der Ansatz von L. Perini (2002ff.) als Bezugspunkt für die weitere Analyse und Arbeit an den zentralen Begriffen gewählt. Als eine der wenigen Autoren berücksichtigt Perini nicht nur alle vier der zu klärenden Begriffe, sondern sie nimmt auch Bezug auf Beispiele aus den Bereichen der computergestützten Mikro- und Systembiologie, was die Abgleichung der Praxisintuitionen im Rahmen dieser Arbeit erleichtert hat. Gleichzeitig hat ihr Ansatz aber auch das Potential, über diese empirischen Bezugskontexte hinauszuweisen dank einer allgemeinen semiotischen Analyse, die einen wichtigen Teil des von ihr verwendeten Begriffsapparates (*frameworks*) ausmacht.

Das große Versäumnis ihres Ansatzes ist es jedoch, dass es Perini nicht gelingt, *genuin bildliche* (nach Goodman *syntaktisch dichte*) Symbolsysteme systematisch so einzubinden, dass *für* ihre funktionale Integrierbarkeit argumentiert werden kann, was jedoch den Intuitionen der Praxis entsprechen würde. Dieses Versäumnis liegt darin begründet, dass Perinis Argumente auf den Begriff der Übersetzbarkeit angewiesen sind, der nicht mit den syntaktischen Merkmalen bildlicher Systeme in Einklang zu bringen ist.

Dass die systematische Integration und leistungsbezogene Beurteilung genuin bildlicher Mittel auf der Grundlage von, beziehungsweise in, Modellen der wissenschaftlichen Repräsentation und besonders der Argumentation nichtsdestoweniger möglich ist, und zwar ohne die Ausgangsbedingungen und Anforderungen an diese Modelle zu stipulieren, das sind die zentralen Thesen dieser Arbeit. Für sie konnte durch die Einbeziehung von weiteren Begriffen, Brückenprinzipien und Bedingungen beziehungsweise Merkmalen der Praxis argumentiert werden. Dazu gehören vor allem die Begriffe der *ad hoc Syntax* und der *ostentativen Spezifikation*, mit deren Hilfe die gewählten Beispielsfälle integriert und begreiflich gemacht werden können.

Im zweiten Teil der Analyse, die den systemspezifischen, technisch oder ökonomisch begründeten Leistungen des Gebrauchs bildlicher Mittel gewidmet ist, konnten weitere Implikationen gewonnen werden für die Revision von Argumenten an den Schnittstellen zwischen der Philosophie und der wissenschaftlichen Visualisierung als Forschungsprogramm. Die Analyse ist in beiden Fällen durch neuartige Praxisbeispiele gestützt, die in diesem Zusammenhang erschlossen und in die Diskussion eingebracht werden konnten, dank der engen Zusammenarbeit mit den beteiligten Wissenschaftlern.

Literature

Achterhuis H. (ed.) (2001). *American Philosophy of Technology. The Empirical Turn*. Bloomington and Indianapolis: Indiana University Press.

Adelmann, R. et al. (2009). *Datenbilder: Zur digitalen Bildpraxis in den Naturwissenschaften*. Bielefeld: transcript.

Alloa, E. (2010). Seeing-As, Seeing-In, Seeing-With: Looking Through Images. In E. Nehmet, R. Heinrich & W. Pichler (eds.), *Image and Imaging in Philosophy, Science, and the Arts*. Proceedings of the 33rd IWS in Kirchberg, 192–204.

Assmann, J. (2009). Ägyptische Bildpraxen und ihre impliziten Theorien. In K. Sachs-Hombach (ed.), *Bildtheorien. Anthropologische und kulturelle Grundlagen des Visualistic Turn*. Frankfurt: Suhrkamp, 74–103.

Auerbach, E. (1967). Figura. In *Gesammelte Aufsätze zur romanischen Philologie*. Bern und München: Francke Verlag, 55–93.

Bach, K. (1970). Part of what a picture is. *British Journal of Aesthetics 10 (2)*, 119–137.

Baigrie, B. (ed.) (1996). *Picturing Knowledge: Historical and Philosophical Problems concerning the Use of Art in Science*. Toronto, Buffalo, London: University of Toronto Press.

Bailer-Jones, D. M. (1999). Tracing the Developement of Models in the Philosophy of Science. In L. Magnani, N. J. Nersessian & P. Thagard (eds.), *Model-Based Reasoning in Scientific Discovery*. New York: Kluwer Academic/Plenum Publishers, 23–41.

Bailer-Jones, D. M. (2002). Models, Metaphors, and Analogies. In P. Machamer & S. Silberstein (eds.), *Guide to the Philosophy of Science*. Oxford: Blackwell, 108–127.

Bailer-Jones, D. M. (2009). *Scientific Models in Philosophy of Science*. Pittsburgh: University of Pittsburgh Press.

Baird, D. (2004). *Thing Knowledge: A Philosophy of Scientific Instruments*. Berkeley: University of California Press.

Barthes, R. (1964). Rhétorique de l'image. *Communication 4*, 40–51.

Barthes, R. (1977). Rhetoric of the Image. In *Image, Music, Text*. Essays selected and translated by S. Heath. London: Fontana/Collins, 32–51.

Bertalanffy, L. v. (1968). Symbolismus und Anthropogenese. In B. Rensch (ed.) *Handgebrauch und Verständigung bei Affen und Frühmenschen.* Bern/Stuttgart: Verlag Hans Huber, 31–149.

Bertin, J. (1974). *Graphische Semiologie. Diagramme Netze Karten.* Translated by G. Jensch, D. Schade & W. Scharfe. Berlin/New York: De Gruyter.

Beyer, M. & Lohhoff, A. (eds.) (2005). *Bild und Erkenntnis: Formen und Funktionen des Bildes in Wissenschaft und Technik.* München: Deutscher Kunstverlag.

Bird, A. & Tobin, E. (2008). Natural Kinds. *The Stanford Encyclopedia of Philosophy.* Edward N. Zalta (ed.). URL <http://plato.stanford.edu/entries/natural-kinds/>

Bisanz, E. (2002). *Malerei als écriture: Semiotische Zugänge zur Abstraktion.* Wiesbaden: Deutscher Universitäts-Verlag.

Boehm, G. (1994). Die Wiederkehr der Bilder. In G. Boehm (ed.), *Was ist ein Bild?* München: Wilhelm Fink.

Boehm, G. (2001). Zwischen Auge und Hand: Bilder als Instrumente der Erkenntnis. In B. Heintz & J. Huber (eds.), *Mit dem Auge denken: Strategien der Sichtbarmachung in wissenschaftlichen und virtuellen Welten.* Zürich: Edition Voldemeer, 43–54.

Boehm, G. (2007). Die ikonische Figuration. In G. Boehm, G. Brandstetter & A. v. Müller (eds.), *Figur und Figuration: Studien zur Wahrnehmung und Wissen.* München: Wilhelm Fink, 33–55.

Boehm, G., Mersmann, B. & Spies, C. (eds.) (2008). *Movens Bild: Zwischen Evidenz und Affekt.* München: Wilhelm Fink.

Boltzmann, L. (1911). Model. *Encyclopedia Britannica.* Cambridge: Cambridge University Press. [1]1902

Brandstetter, G. (2007). SchnittFiguren. Intersektionen von Bild und Tanz. In G. Boehm, G. Brandstetter & A. v. Müller (eds.), *Figur und Figuration: Studien zur Wahrnehmung und Wissen.* München: Wilhelm Fink, 13–33.

Bredekamp, H. (2000). Gazing Hands and Blind Spots: Galileo as Draftsman. *Science in Context 13 (3–4)*, 423–462.

Bredekamp, H., Bruhn M. & Werner G. (eds.) (2003ff.). *Bildwelten des Wissens. Kunsthistorisches Jahrbuch für Bildkritik.* Berlin: Akademie-Verlag.

Bredekamp, H. (2008). Bildbeschreibungen. Eine Stilgeschichte technischer Bilder? Ein Interview mit Horst Bredekamp. In H. Bredekamp, H., Schneider, B. & Dünkel, V. (eds.), *Das Technische Bild: Kompendium zu einer Stilgeschichte der Bilder.* Berlin: Akademie-Verlag, 36–48.

Bredekamp, H. (2010). *Theorie des Bildakts.* Frankfurter Adorno-Vorlesungen 2007, Frankfurt: Suhrkamp.

Breidbach, O. (2005). *Bilder des Wissens: Zur Kulturgeschichte der wissenschaftlichen Wahrnehmung*. Wilhelm Fink, München.

Bridgman, P. W. (1927). *The Logic of Modern Physics*. New York: Macmillan.

Brodlie, K. et al. (eds.) (1992). *Scientific Visualisation: Techniques and Applications*. Berlin/Heidelberg/New York: Springer.

Bungartz, H.-J., Griebel, M. & Zenger, C. (2002). *Einführung in die Computergraphik: Grundlagen, Geometrische Modellierung, Algorithmen*. Braunschweig/Wiesbaden: vieweg.

Cappelen, H. & Lepore, E. (2012). Quotation. *The Stanford Encyclopedia of Philosophy*. Edward N. Zalta (ed.). URL <http://plato.stanford.edu/entries/quotation/>

Carnap, R. (1928). *Der logische Aufbau der Welt*. Berlin-Schlachtensee: Weltkreis Verlag.

Carnap, R. (1931). Überwindung der Metaphysik durch die logische Analyse der Sprache. *Erkenntnis* 2, 219–241.

Carnap, R. (1937). *The Logical Syntax of Language*. New York: Humanities.

Carrier, M. (2006). *Wissenschaftstheorie zur Einführung*. Hamburg: Junius

Carruthers, M. (1998). *The Craft of Thought: Mediation, Rhetoric, and the Making of Images, 400–1200*. Cambridge: Cambridge University Press.

Carruthers, M. & Ziolowski, J. M. (eds.) (2002). *The Medieval Craft of Memory: an Anthology of Texts and Pictures*. Philadelphia: University of Pennsylvania Press.

Cartier, P. (1997). The Continuing Silence of Bourbaki (Interview). *The Mathematical Intelligencer*. New York: Springer.

Cartwright, N. (1983). *How the Laws of Physics Lie*. Oxford: Oxford University Press.

Cartwright, N. (1999). *The Dappled World. A Study of the Boundaries of Science*. Cambridge: Cambridge University Press.

Carusi, A. (2008). Scientific visualization and aesthetic grounds for trust. *Ethics and Information Technology 10*, 243–254.

Carusi, A. (2012). Making the Visual Visible in Philosophy of Science. *Spontaneous Generations: A Journal for the History and Philosophy of Science* 6, 106–114.

Cassirer, E. (2009). *Schriften zur Philosophie der symbolischen Formen*. Edited by M. Lauschke. Hamburg: Felix Meiner. [1]1923ff.

Chang, H. (2009). Operationalism. *The Stanford Encyclopedia of Philosophy*. Edward N. Zalta (ed.). URL <http://plato.stanford.edu/entries/operationalism/#CriOpe>

Chen, M. et al. (2009). Data, Information, and Knowledge in Visualization. *Computer Graphics and Applications (Visualisation Viewpoints) (Jan/Feb)*, 12–19.

Collins, H. (2009). Social Construction of Science. In J. K. B. Olsen, S. A. Pedersen & V. F. Hendricks (eds.), *A Companion to the Philosophy of Technology*. Oxford: Blackwell, 84–87.

Cummins, R. (1975). Functional Analysis. *The Journal of Philosophy 72 (20)*, 741–765.

Danto, A. C. (1994). Abbildung und Beschreibung. In G. Boehm (ed.), *Was ist ein Bild?* München: Wilhelm Fink, 125–147.

Darwin, C. (1872). *The Expression of the Emotions in Man and Animals*. London: John Murray.

Daston, L. & Galison, P. (1992). The Image of Objectivity. *Representations 0 (40)*, 81–128.

Daston, L. & Galison, P. (2007). *Objectivity*. Cambridge MA, Zone Books.

De Regt, H. W. & Dieks, D. (2005), A Contextual Approach To Scientific Understanding. *Synthese 144 (1)*, 137–170.

Descartes, R. (1954). *The Geometry of René Descartes*. Translated by D. E. Smith & M. L. Latham. New York: Dover Publications. [1]1637

Downes, S. M. (2012). How Much Work Do Scientific Images Do? *Spontaneous Generations: A journal for the History and Philosophy of Science 6(1)*, 115–130.

Duhem, P. (1998). *Ziel und Struktur der physikalischen Theorien.* Edited by L. Schäfer. Hamburg: Felix Meiner. [1]1906

Dummett, M. A. E. (1993). *Origins of Analytical Philosophy*. London: Duckworth.

Eco, U. (1976). *A Theory of Semiotics*. Bloomington: Indiana University Press.

Elgin, C. Z. (1983). *With Reference to Reference*. Indianapolis: Hackett Publishing Company.

Elkins, J. (2003). *Visual Studies: A Skeptical Introduction.* New York/London: Routledge.

Elkins, J. (2007). *Visual Practices Across the University*. München: Wilhelm Fink.

Elkins, J. (2008). Introduction: The Concept of Visual Literacy, and Its Limitations. In J. Elkins (ed.), *Visual Literacy*. New York/London: Routledge, 1–10.

Elkins, J. (2010). Visual Practices Across the University. In R. Frigg & M. C. Hunter (eds.), *Beyond Mimesis and Convention: Representation in Art and Science.* Dordrecht: Springer, 169–193.

Elkins, J. (2011). Reasons Why it Is Not Possible to List Concepts of the Image. *Rheinsprung 11 – Zeitschrift für Bildkritik 01*, 13–20.

Elpidorou, A. (2010). Imagination in Non-Representational Painting. In J. Webber (ed.), *Reading Sartre: On Phenomenology and Existentialism*. London: Routledge, 15–31.

Ertl, T. (2009). Visualization [Lecture Slides] – Summer Term 2010. URL <http://www.vis.uni-stuttgart.de/lehre/archiv/> [access on demand].

Falk, M., Klann, M., Reuss, M. & Ertl, Th. (2009). Visualization of Signal Transduction Processes in the Crowded Environment of the Cell. *IEEE Pacific Visualization Symposium (PacificVis 2009)*, 169–176.

Falkenburg, B. (1997). Modelle, Korrespondenz und Vereinheitlichung in der Physik. *Dialektik 1*, 28–42.

Feyerabend, P. (2010). *Against Method*. London/New York: Verso. [1]1975

Fischer, M. (1997). Schrift als Notation. In P. Koch & S. Krämer (eds.), *Schrift, Medien, Kognition*. Tübingen: Stauffenburg Verlag, 83–105.

Fleck, L. (1980). *Entstehung und Entwicklung einer wissenschaftlichen Tatsache*. Edited by Schäfer, L. & Schnelle, T. Frankfurt: Suhrkamp.

Fleck, L. (1983). Schauen, sehen, wissen. In L. Fleck, *Erfahrung und Tatsache*. Edited by Schäfer, L. & Schnelle. Frankfurt: Suhrkamp, 147–175.

Fraassen, B. C. v. (1980). *The Scientific Image*. Oxford University Press.

Fraassen, B. C. v. (1991). *Quantum Mechanics: An Empiricist View*. Clarendon Press.

Fraassen, B. C. v. (2002). *The Empirical Stance*. Yale: Yale University Press.

Fraassen, B. C. v. (2008). *Scientific Representation: Paradoxes of Perspective*. Oxford: Clarendon Press.

Frankel, F. (2002). *Envisioning science: the design and craft of the science image*. Cambridge: MIT Press.

Frege, G. (1998). *Begriffsschrift und andere Aufsätze*. Hildesheim: Olm. [1]1879

French, S. & Ladyman, J. (1999). Reinflating the Semantic Approach. *International Studies in the Philosophy of Science 13 (2)*, 103–121.

Friedhoff, R. & Kiely, T. (1990). The eye of the beholder. *Computer Graphics World 13(8)*, 47–56.

Frigg, R. (2003). *Re-Presenting Scientific Representation* (Diss.). London: LSE

Frigg, R. (2006). Scientific Representation and the Semantic View of Theories. *Theoria 55*, 49–65.

Frigg, R. & Hartmann, S. (2006). Models in Science. *The Stanford Encyclopedia of Philosophy*. Edward N. Zalta (ed.). URL <http://plato.stanford.edu/entries/models-science/>

Frigg, R. (2010). Fiction and Scientific Representation In R. Frigg & M. C. Hunter (eds.), *Beyond Mimesis and Convention: Representation in Art and Science*. Dordrecht: Springer, 97–138.

Gabriel, G. (2000). Kontinentales Erbe und analytische Methode. Nelson Goodman und die Tradition. *Erkenntnis 52 (2)*, 185–198.

Gabriel, G. & Zantwijk, T. v. (2001). Urteil. In J. Ritter, K. Gründer & G. Gabriel (eds.), *Historisches Wörterbuch der Philosophie (HWPh)*. Basel: Schwabe, 436–444.

Gadamer, H.-G. (1972). *Wahrheit und Methode*. Tübingen: Mohr.

Gall, A. (ed.) (2007). *Konstruieren, Kommunizieren, Präsentieren: Bilder von Wissenschaft und Technik*. Göttingen: Wallstein Verlag.

Gallagher, R. S. (1995). *Computer Visualisation: Graphics Techniques for Scientific and Engineering Analysis*. Boca Raton: CRC Press.

Gaycken, S. (2005). Was ist technisches Wissen? *Proceedings 22C3*, 1–18.

Gebauer, G. (2005). Zwei Bilderwelten. In Ch. Bezzel (ed.), *Sagen und Zeigen: Wittgensteins „Tractatus", Sprache und Kunst*. Berlin: ParErga, 31–55.

Giaquinto, M. (2007). *Visual Thinking in Mathematics*. Oxford: Oxford University Press.

Giere, R. (1988). *Explaining Science*. Chicago: University of Chicago Press.

Giere, R. N. (1996), Visual Models and Scientific Judgement. In B. Baigrie (ed.), *Picturing Knowledge: Historical and Philosophical Problems Concerning the Use of Art in Science*. Toronto/Buffalo/London: University of Toronto Press: 269–303.

Giere, R. (1999). *Science Without Laws*. Chicago: University of Chicago Press.

Giere, R. (2004). How Models Are Used to Represent Reality. *Philosophy of Science 71*, 742–752.

Giere, R. N. (2006). *Scientific Perspectivism*, University of Chicago Press.

Gombrich, E. H. (1977). *Art and Illusion: A study in the psychology of pictorial representation*. Oxford: Phaidon Press.

Goodman, N. (1976). *Languages of Art: An Approach To A Theory Of Symbols*. Indianapolis: Hackett Publishing Company.

Goodman, N. & Lüdeking, K. (2005). Gewissheit ist etwas ganz und gar Absurdes (Interview). In J. Steinbrenner, O. Scholz & G. Ernst (eds), *Symbole, Systeme, Welten*. Heidelberg: Synchron. Wissenschaftsverlag der Autoren, 261–271.

Goodwin, W. (2009). Visual Representations in Science. *Philosophy of Science 76 (3)*, 372–390.

Gramelsberger, G. (2000). *Semiotik und Simulation: Fortführung der Schrift ins Dynamische. Entwurf einer Symboltheorie der numerischen Simulation und ihrer Visualisierung* (Diss.) FU Berlin.

Gramelsberger, G. (2012). Schrift auf den Punkt gebracht – Extrapolation, Rekursion, Simulation. In S. Krämer, E. Cancik-Kirschbaum & R. Totzke (eds.), *Schriftbildlichkeit: Wahrnehmbarkeit, Materialität und Operativität von Notationen*. Berlin: Akademie Verlag, 389–400.

Grandin, T. (2006). *Thinking in Pictures: And Other Reports from my Life with Autism*. London: Bloomsbury.

Gugerli, D. (ed.) (2002). *Ganz normale Bilder: historische Beiträge zur Herstellung von Selbstverständlichkeit*. Zürich: Chronos.

Habbe, K. A. (1996). Das methodische Instrument „Beschreibung“ in der Geographie. In R. Kötter & R. Inhetveen (eds.), *Betrachten – Beobachten – Beschreiben: Beschreibungen in Kultur- und Naturwissenschaften*. München: Wilhelm Fink, 15–49.

Hager, F.-P. (1971). Episteme. In J. Ritter, K. Gründer & G. Gabriel (eds.), *Historisches Wörterbuch der Philosophie (HWPh)* Basel: Schwabe, 588–593.

Hahn, H. (1933). Logik, Mathematik und Naturerkennen. In O. Neurath; R. Carnap & H. Hahn (eds.), *Einheitswissenschaft 2*. Wien: Gerold & Co., 5–33.

Hanson, N. R. (1958). *Patterns of Discovery*. Cambridge: University of Cambridge Press.

Harré, R. (1970). *The Principles of Scientifc Thinking*. New York: Macmillan.

Hartmann, F. & Bauer, E. K. (2006). *Bildersprache: Otto Neurath: Visualisierungen*. Wien: Facultas.

Heintz, B. & Huber, J. (eds.) (2001). *Mit dem Auge denken: Strategien der Sichtbarmachung in wissenschaftlichen und virtuellen Welten*. Zürich: Edition Voldemeer.

Hempel, C. G. (1965). The Logic of Functional Analysis. In C. G. Hempel (ed.), *Aspects of Scientific Explanation*. New York: Free Press, 271–307. [1]1959

Hentschel, K. (2000). Drawing, engraving, photographing, plotting, printing: Historical studies of visual representations, particularly in astronomy. In K. Hentschel & A. D. Wittmann (eds.), *The Role of Visual Representations in Astronomy: History and Research Practice*. Frankfurt: Harri Deutsch, 11–43.

Hertz, H. (1894). Die Prinzipien der Mechanik. In H. Hertz (ed.), *Gesammelte Werke, Band 3*, Leipzig: Johann Ambrosius Barth.

Hesse, M. (1966). *Models and Analogies in Science*. Indiana: University of Notre Dame Press. [1]1963

Hesse, M. (1980). *Revolutions and Reconstructions in the Philosophy of Science*. Brighton: The Harvester Press.

Heßler, M. (ed.) (2004). *Explorationsstudie im Rahmen der BMBF- Förderintitiave „Wissen für Entscheidungsprozesse“ zum Thema Visualisierungen in der Wissenschaftskommunikation*. Aachen: BMBF Förderinitiative.

Heßler, Martina (ed.) (2006). *Konstruierte Sichtbarkeiten. Wissenschafts- und Technikbilder seit der Frühen Neuzeit*. München: Wilhelm Fink.

Hoffmann, S. (2002). *Die Geschichte des Medienbegriffs*. Hamburg: Meiner.

Hopkins, R. (1998). *Picture, Image and Experience: A Philosophical Inquiry*. Cambridge: Cambridge University Press.

Hosten, N. (2010). Denken in Röntgenbildern. In U. Nortmann & Ch. Wagner (eds.), *In Bildern Denken?* München: Wilhelm Fink, 141–146.

Hoyningen-Huene, P. (2002). On the Varieties of the Distinction between the Context of Discovery and the Context of Justification. in J. Schickore & F. Steinle, (eds.), *Revisiting Discovery and Justification*. Berlin: Preprint 211 of the Max Planck Institute für Wissenschaftsgeschichte, 11–16.

Hughes, R. I. G. (1997). Models and Representation. *Philosophy of Science 64 (4)*, 325–336.

Hughes, J., Kroes, P. & Zwart, S. (2007). A semantics for means-end relations. *Synthese 158*, 207–231.

Hunter, M. C. (2010). Experiment, Theory, Representation: Robert Hooke's Material Models. In R. Frigg & M. C. Hunter (eds.), *Beyond Mimesis and Convention: Representation in Art and Science*. Dordrecht: Springer, 169–193.

Husserl, E. (2006). Phantasie und Bildbewusstsein. In E. Marbach (ed.), *Husserliana* (Vol. XXIII). Hamburg: Felix Meiner. [1]1904/05

Hutten, E. H. (1956). *The Language of Modern Physics*. London: George Allen & Unwin LTD.

Idhe, D. (2009). Imaging Technologies. In J. K. B. Olsen, S. A. Pedersen & V. F. Hendricks (eds.), *A Companion to the Philosophy of Technology*. Oxford: Blackwell, 205–209.

Iser, W. (1975). Der Lesevorgang. In R. Warning (ed.), *Rezeptionsästhetik: Theorie und Praxis*. München: Wilhelm Fink, 253–277.

Ivins (Jr.), W. M. (1969). *Prints and Visual Communication*. Cambridge: MIT Press.

Jonas, H. (1961). Die Freiheit des Bildens – Homo pictor und die differentia des Menschen. *Zeitschrift für philosophische Forschung 15*, 161–176.

Kapust, A. (2009). Phänomenologische Bildpositionen. In K. Sachs-Hombach (ed.), *Bildtheorien: Anthropologische und kulturelle Grundlagen des Visualistic Turn*. Frankfurt: Suhrkamp, 255–284.

Kemp, M. (1990). Taking it on Trust: form and meaning in naturalistic representation. *Archives of Natural History 17 (2)*, 127–188.

Kemp, M. (2000). *Visualizations: The Nature Book of Art and Science*. Oxford: Oxford University Press.

Kepler, J. (1938). Mysterium Cosmographicum. De Stella Nova. *In Gesammelte Werke* (Vol. 1). Edited by M. Caspar. München: C.H. Becksche Verlagsbuchhandlung. [1]1596

Kirschenmann, P. P. (2010). On the Truth of Pictures in Science. In E. Nehmet, R. Heinrich & W. Pichler (eds.), *Image and Imaging in Philosophy, Science, and the Arts*. Proceedings of the 33rd IWS in Kirchberg, 152–154.

Kjörup, S. (1978). Pictorial Speech Acts. *Erkenntnis 12*, 55–71.

Knorr-Cetina, K. & Amann, K. (1990). Image Dissection in Natural Scientific Inquiry. *Science, Technology and Human Values 15 (3)*, 259–283.

Knorr-Cetina, K. (2001). Objectual Practice. In Th. R. Schatzki; K. Knorr-Cetina & E. v. Savigny (eds.), *The Practice Turn in Contemporary Theory*. London: Routledge, 175–188.

Knuuttila, T. & Voutilainen, A. (2002). A Parser as an Epistemic Artifact: a Material View on Models. *Philosophy of Science 70 (5)*, 1484–1495.

Koch, P. (1997). Graphé. Ihre Entwicklung zur Schrift, zum Kalkül und zur Liste. In P. Koch & S. Krämer (eds.), *Schrift, Medien, Kognition*. Tübingen: Stauffenburg, 43–83.

Krämer, S. (1988). *Symbolische Maschinen: Die Idee der Formalisierung in geschichtlichem Abriss*. Darmstadt: Wissenschaftliche Buchgesellschaft.

Krämer, S. (ed.) (2004). *Performativität und Medialität*. München: Wilhelm Fink.

Krämer, S. (2012). Punkt, Strich, Fläche. In S. Krämer, E. Cancik-Kirschbaum & R. Totzke (eds.), *Schriftbildlichkeit: Wahrnehmbarkeit, Materialität und Operativität von Notationen*. Berlin: Akademieverlag, 79–100.

Krämer, S. & Totzke, R. (2012). Einleitung. In S. Krämer, E. Cancik-Kirschbaum & R. Totzke (eds.), *Schriftbildlichkeit: Wahrnehmbarkeit, Materialität und Operativität von Notationen*. Berlin: Akademieverlag, 13–39.

Kroes, P., Meijers A. & C. Mitcham (eds.) (2001). *The Empirical Turn in the Philosophy of Technology*. Oxford: Elsevier Science.

Kuhn, T. (1996). *The Structure of Scientific Revolutions*. Chicago/London: The University of Chicago Press. [1]1962

Kötter, R. (2007). Abbildung, Verbildlichung, Veranschaulichung: Zu Grundfunktionen des Bildes in den Wissenschaften. In F. Stahnisch & H. Bauer (eds.), *Bild und Gestalt: Wie formen Medienpraktiken das Wissen in Medizin und Humanwissenschaften?* Hamburg: Lit, 161–181.

Lagrange, J. (1997). *Analytical Mechanics*. Edited by A. C. Boissonnade & V. N. Vagliente. Boston: Kluwer Academic. [1]1811

Lagrange, J. L. (1853). *Mécanique Analytique*. Edited by M. J. Bertrand. Paris: Mallet-Bachelier. [1]1818

Lahsen, M. (2005). Seductive Simulations? Uncertainty Distribution Around Climate Models. *Social Studies of Studies 35*, 895–922.

Lakatos, I. (1978). *The Methodology of Scientific Research Programmes: Philosophical Papers Volume 1*. Cambridge: Cambridge University Press.

Langer, S. K. (1979). *Philosophy in a New Key: A Study in the Symbolism of Reason, Rite and Art*. Cambridge: Harvard University Press.

Latour, B. & Woolgar, S. (1986), Laboratory Life: The Construction of Scientific Facts, Princeton University Press, Princeton.

Latour, B. (1987). *Science in Action: How to follow scientists and engineers through society*. Cambridge: Harvard University Press.

Latour, B. (1996). Arbeit mit Bildern oder: Die Umverteilung der wissenschaftlichen Intelligenz. In B. Latour, *Der Berliner Schlüssel: Erkundungen eines Liebhabers der Wissenschaften*. Berlin: Akademie Verlag, 159–190.

Laudan, L. (1977). *Progress and its problems: toward a theory of scientific growth*. Berkely: University of California Press.

Lechtermann, C. & Morsch C. (2004). *Kunst der Bewegung. Kinästhetische Wahrnehmung und Probehandeln in virtuellen Welten*. Frankfurt: Peter Lang.

Lessing, G. (1964). *Laokoon oder Über die Grenzen der Malerei und Poesie*. Stuttgart: Reclam.

Luft, A. L. & Kötter, R. (1994). *Informatik – eine moderne Wissenstechnik*. Mannheim/Leipzig/Wien/Zürich: BI Wissenschaftsverlag.

Lynch, M. (1985). Art and artifact in laboratory science: A study of shop work and shop talk in a research laboratory. London/Boston: Routledge & Kegan Paul.

Lynch, M. (1988). The externalized retina: Selection and mathematization in the visual documentation of objects in the life sciences. *Human Studies 11*, 201–234.

Lynch, M. & Woolgar S. (1990). Introduction: Sociological orientation to representational practice in science. In M. Lynch & S. Woolgar (eds.), *Representation in Scientific Practice*. Cambridge/London: MIT Press, 1–19.

MacFarlane, J. (2000). *What does it mean to say that logic is formal?* (Diss.). University of Pittsburgh.

Marey, E. J. (1878). *La méthode graphique dans les sciences expérimentales et principalément en physiologie et en medicine*. Paris: G. Masson.

Maxwell, J. C. (1965). On Faraday's Lines of Force. In W. D. Niven (ed.), *The Scientific Papers of James Clerk Maxwell*. New York: Dover Publications. [1]1855

Maynard, P. (2009). drawing, painting, and printmaking. In S. Davies et al. (eds.), *A Companion to Aesthetics*. Oxford: Blackwell, 82–85.

Mersch, D. (2006). Wittgensteins Bilddenken. *DZPhil 54*, 6, 925–942.

Mersch, D. (2006a). *Medientheorien zur Einführung*. Junius.

Mitchell, W. J. T. (1984). What Is an Image? *New Literary History 15 (3)*, 503–537.

Mitchell, W. J. T. (1988). Wittgenstein's Imagery and What It Tells Us. *New Literary History 19 (2)*, 361–370.

Mitchell, W. J. T. (1994). The Pictorial Turn. In W. J. T. Mitchell (ed.), *Picture theory: essays on verbal and visual representation*, University of Chicago Press, 11–35.

Mitchell, W. J. T. (2008), Visual Literacy or Literary Visualcy? In J. Elkins (ed.), *Visual Literacy*. New York: Routledge, 11–29.

Morgan, M. S. & Morrison, M. (1999). *Models as Mediators*. Cambrigde: Cambridge University Press.

Moulines, C. U. (2002). Introduction: Structuralism as a Program for Modeling Theoretical Science. *Synthese 130*, 1–11.

Müller, O. (2003). John Stuart Mills Argument für den Utilitarismus: Ein plausibler Weg zwischen Metaphysik und Nihilismus? In U. Meixner & A. Newen (eds.), *Geschichte der Ethik: Jahrbuch 6 für Philosophiegeschichte und Logische Analyse*. Paderborn: Mentis, 167–191.

Nagel, E. (1961). *The Structure of Science: Problems in the Logic of Scientific Explanation*. New York: Harcourt.

Neurath, O. (1933). *Einheitswissenschaft und Psychologie*. Wien: Gerold & Co.

Neurath, O. (1991). *Gesammelte bildpädagogische Schriften*. Edited by R. Haller. Wien: Hoelder-Pichler-Tempsky. 1933ff.

Nielson, G. M., Hagen, H. & Müller, H. (eds.) (1997). *Scientific Visualization: Overviews, Methodologies, and Techniques*. Los Alamitos: IEEE Computer Society.

Panofsky, E. (1974). *Aufsätze zu Grundfragen der Kunstwissenschaft*. Edited by H. Oberer & E. Verheyen, E. Berlin: Verlag Bruno Hessling. 1932ff.

Panofsky, E. (1932). Zum Problem der Beschreibung und Inhaltsdeutung von Werken der bildenden Kunst. *Logos 21*, 103–119.

Panofsky, E. (1956). Galileo as a Critic of the Arts: Aesthetic Attitude and Scientific Thought. *Isis 47 (1)*, 3–15.

Parker, W. S. (2009). Confirmation and Adequacy-for-Purpose in Climate Modelling. *Proceedings of the Aristotelian Society Supplementary LXXXIII*, 233–249.

Peirce, C. S. (1955). *Philosophical Writings of Peirce*. Edited by J. Buchler. New York: Dover Publications.

Pelz, H. (2005). *Linguistik: eine Einführung*. Hamburg: Hoffman und Campe.

Perini, L. (2002). *Visual Representations and Scientific Knowledge* (Diss.). San Diego: University of California.

Perini, L. (2005). The Truth in Pictures. *Philosophy of Science 72*, 262–285.

Perini, L. (2005a). Visual Representation and Confirmation. *Philosophy of Science 72*, 913–926.

Perini, L. (2005b). Explanation in two dimensions: diagrams and biological explanation. *Biology and Philosophy 20*, 257–269.

Perini, L. (2010). Scientific Representation and the Semiotics of Pictures. In P. D. Magnus & J. Busch (eds.), *New waves in philosophy of science*. New York: Macmillan, 131–154.

Perini, L. (2012). Truth-bearers or Truth-makers? *Spontaneous Generations: A Journal for the History and Philosophy of Science 6*, 142–147.

Perini, L. (2012a). Depiction, Detection, and the Epistemic Value of Photography. *The Journal of Aesthetics and Art Criticism 70*, 151–160.

Pickering, A. (1992). From Science as Knowledge to Science as Practice. In A. Pickering (ed.), *Science as Practice and Culture*. Chicago/London: The University of Chicago Press, 1–29.

Pichler, A. (2010). Wie man ein Luftgebäude zerstört, oder: Die Bedeutung von Sagen und Zeigen für Nietzsches und Wittgensteins Praxis philosophischer Kritik. In E. Nehmet, R. Heinrich & W. Pichler (eds.), *Image and Imaging in Philosophy, Science, and the Arts*. Proceedings of the 33rd IWS in Kirchberg, 192–204

Platon (2011). *Politeia* [370 BC]. Edited by O. Höffe. Berlin: Akademie-Verlag.

Polanyi, M. (1958). *Personal Knowledge: towards a post-critical philosophy*. London/Boston: Routledge & Kegan Paul.

Pompe, U. (2011). *Perception and Cognition. The analysis of object recognition* (Diss.). Paderborn: Mentis.

Popper, K. (1959). *The Logic of Scientific Discovery*. London: Hutchinson.

Psillos, S. (1999). *Scientific Realism. How Science Tracks Truth*. Oxford: Routledge.

Rayner, K. (1998). Eye Movements in Reading and Information Processing: 20 Years of Research. *Psychological Bulletin 124 (3)*, 372–422.

Reichenbach, H. (1938). *Experience and Prediction: An Analysis of the Foundations and Structure of Knowledge*. Chicago/London: The University of Chicago Press.

Rheinberger, H.-J. (1992). *Experiment-Differenz-Schrift: Zur Geschichte epistemischer Dinge*. Berlin: Basilisken-Presse.

Rheinberger, H.-J. (2001). Objekt und Repräsentation. In B. Heintz & J. Huber (eds.), *Mit dem Auge denken: Strategien der Sichtbarmachung in wissenschaftlichen und virtuellen Welten*. Zürich: Edition Voldemeer, 55–61.

Rheinberger, H.-J. (2005). Epistemologica: Präparate. In A. te Heesen & P. Lutz (eds.), *Dingwelten: Das Museum als Erkenntnisort*. Köln/Weimar/Wien: Böhlau, 65–77.

Rheinsprung 11 (2011ff.). *Zeitschrift für Bildkritik*. Basel: eikones NFS Bildkritik/ NCCR Iconic Criticism.

Richter, M. (2011). Virtual Reality and Virtual Actuality: Remarks on the Use of Technical Terms in Philosophy of Virtuality. In C. Ess & M. Thorseth (eds.), *Trust and Virtual Worlds: Contemporary Perspectives*. Frankfurt: Peter Lang, 31–44.
Richtmeyer, U. (2009). Logik und Aisthesis: – Wittgenstein über Negationen, Variablen und Hypothesen im Bild. In M. Heßler & D. Mersch (eds.), *Logik des Bildlichen: Zur Kritik der ikonischen Vernunft*. Bielefeld: transcript, 139–163.
Robin, H. (1992). *The scientific image: from cave to computer*. New York: Abrams.
Rorty, R. (1967). *The Linguistic Turn: Essays in philosophical method*. Chicago/London: University of Chicago Press.
Rudwick, M. J. S. (1976). The Emergence of a Visual Language for Geological Science 1760–1840. *History of Science 14*, 149–195.
Sachs-Hombach, K. (2003). *Das Bild als kommunikatives Medium: Elemente einer allgemeinen Bildwissenschaft* Köln: Herbert von Halem.
Sachs-Hombach, K. (ed.) (2009). *Bildtheorien: Anthropologische und kulturelle Grundlagen des Visualistic Turn*. Frankfurt: Suhrkamp.
Schich, M. (2009). *Rezeption und Tradierung als komplexes Netzwerk: Der CENSUS und visuelle Dokumente zu den Thermen in Rom* (Diss.). München: Biering & Brinkmann.
Schier, F. (1986). *Deeper Into Pictures: An Essay on Pictorial Representation*. Cambridge: Cambridge University Press.
Schirra, J. R. J. (1999). Syntaktische Dichte oder Kontinuität – Ein mathematischer Aspekt der Visualistik. In K. Sachs-Hombach & K. Rehkämper (eds.), *Bildgrammatik: Interdisziplinäre Forschungen*. Magdeburg: Scriptum Verlag, 103–123.
Schittler, D., Hasenauer, J., Allgöwer F. & Waldherr, S. (2010). Cell differentiation modeled via a coupled two-switch regulatory network. *CHAOS 20, 045121*, 1–9.
Schmidt-Burkhardt, A. (2009). Wissen als Bild. Zur diagrammatischen Kunst-geschichte. In M. Heßler & D. Mersch (eds.), *Logik des Bildlichen*. Bielefeld: transcript, 163–188.
Scholz, O. (1999). Mein teurer Freund, ich rat' euch drum/Zuerst Collegium Syntacticum – Das Meisterargument in der Bildtheorie. In K. Sachs-Hombach (ed.), *Bildgrammatik*. Scriptum Verlag, 33–47.
Scholz, O. (2004). *Bild, Darstellung, Zeichen*. Frankfurt: Vittorio Klostermann.
Schumann, H. & Müller, W. (2000). *Visualisierung: Grundlagen und allgemeine Methoden*. Berlin/Heidelberg/New York: Springer.
Seja, S. (2009). *Handlungstheorien des Bildes*. Köln: Herbert von Halem.
Senechal, M. (1998). The Continuing Silence of Bourbaki – An Interview with Pierre Cartier. *The Mathematical Intelligencer (Springer New York) 19(4)*, 22–28.

Shapere, D. (1984). *Reason and the search for knowledge: investigation in the philosophy of science*. Dordrecht: Reidel.

Soraker, J. H. (2010). *The Value of Virtual Worlds and Entities – A Philosophical Analysis of Virtual Worlds and Their Potential Impact on Well-Being* (Diss.). Enschede: Ipskamp.

Soraker, J. H. (2011). Virtual Entities, Environments, Worlds and Reality: Suggested Definitions and Taxonomy. In C. Ess & M. Thorseth (eds.), *Trust and Virtual Worlds: Contemporary Perspectives*. Frankfurt: Peter Lang, 44–75.

Spence, R. (2007). *Information Visualization: Design for Interaction*. Englewood cliffs: Prentice Hall.

Stegmüller, W. (1969). *Hauptströmungen der Gegenwartsphilosophie*. Stuttgart: Kröner.

Stegmüller, W. (1976). *The Structure and Dynamics of Theories*. Berlin: Springer.

Steinbrenner, J. (2009). Bildtheorien der analytischen Tradition. In K. Sachs-Hombach (ed.), *Bildtheorien. Anthropologische und kulturelle Grundlagen des Visualistic Turn*. Frankfurt: Suhrkamp, 248–315.

Steinbrenner, J. (2005). Zitat. In J. Ritter, K. Gründer & G. Gabriel (eds.), *Historisches Wörterbuch der Philosophie (HWPh)*. Basel: Schwabe, 1355–1357.

Steiner, H. G. (1976). Isomorphie. In J. Ritter, K. Gründer & G. Gabriel (eds.), *Historisches Wörterbuch der Philosophie (HWPh)*. Basel: Schwabe, 627–628.

Stenius, E. (1971). Bild, logisches. In J. Ritter, K. Gründer & G. Gabriel (eds.), *Historisches Wörterbuch der Philosophie (HWPh)*. Basel: Schwabe, 920–921.

Stern, D. G. (2003). The Practical Turn. In S. P. Turner & P. A. Roth (eds.), *The Blackwell Guide to the Philosophy of the Social Sciences*. Oxford: Blackwell, 185–207.

Summerfield, D. M. (1996). Fitting versus tracking: Wittgenstein on representation. In H. Sluga & D. G. Stern (eds.), *The Cambridge Companion to Wittgenstein*. Cambridge: Cambridge University Press, 100–139.

Suppe, F. (1989). *The Semantic Conception of Theories and Scientific Realism*. Chicago/London: University of Chicago Press.

Suppes, P. (1969). *Studies in the Methodology and Foundations of Science: selected papers from 1951 to 1969*. Dordrecht: Reidel.

Tarski, A. (1956). *The Concept of Truth in Formalized Languages. Logic, Semantics, Mathematics*. Oxford: Clarendon Press. [1]1933

Tarski, A. (1995). *Introduction to logic and to the methodology of deductive sciences*. New York: Dover Publications. [1]1946

Thomson, W. (1987). Lecture XX. In R. Kargon & P. Aichinstein (eds.), *Kelvin's Baltimore Lectures and Modern Theoretical Physics: Historical and Philosophical Perspectives*. Cambridge: MIT Press. [1]1884

Thomson-Jones, M. (2006). Models and the Semantic View. *Philosophy of Science 73(5)*, 524–535.

Topper, D. (1996). Towards an Epistemology of Scientific Illustration. In B. Baigrie (ed.), *Picturing Knowledge: Historical and Philosophical Problems Concerning the Use of Art in Science*. Toronto/Buffalo/London: University of Toronto Press, 215–250.

Toulmin, S. (1972). *Human Understanding: The Collective Use and Evolution of Concepts*. Princeton: Princeton University Press.

Vogelgsang, T. (2009). *Von Linien und Kurven: Johann Heinrich Lambert und der Graph der magnetischen Abweichung* (Magister Thesis). HU Berlin.

Waddington, C. H. (1957). *The Strategy of the Genes: A Discussion of some Aspects of Theoretical Biology*. London: Allen & Unwin.

Walton, K. L. (1990). *Mimesis as Make-Believe: On the Foundations of the Representational Arts*. Cambridge: Cambridge University Press.

Walton, K. L. (1976). Points of View in Narrative and Depictive Representation. *Noûs 10 (1)*, 49–61.

Walton, K. L. (1973). Pictures and Make-Believe. *The Philosophical Review 82 (3)*, 283–319.

Ward, M., Grinstein, G. G. & Keim, D. (2010). *Interactive Data Visualization: Foundations, Techniques, and Applications*. London: Taylor & Francis.

Warnke, P. (2002). *Computersimulation und Intervention: Eine Methode der Technikentwicklung als Vermittlungsinstrument soziotechnischer Umordnungen* (Diss.). Technische Universiät Darmstadt.

Wedekind, H., Görz, G., Kötter, R. & Inhetveen, R. (1998). Modellierung, Simulation, Visualisierung: Zu aktuellen Aufgaben der Informatik. *Informatik-Spektrum 21*, 265–272.

Wimsatt, W. C. (1990). Taming the Dimensions-Visualizations in Science. *PSA: Proceedings of the Biennial Meeting of the Philosophy of Science Association 2*, 111–135.

Winkler, H. (2004). How to do things with words, signs, machines. In S. Krämer (ed.), *Performativität, Medien, Praxen, Computer*. München: Wilhelm Fink.

Winsberg, E. (1999). Sanctioning Models: The Epistemology of Simulation. *Science in Context 12*, 275–292.

Wittgenstein, L. (1961), *Notebooks 1914–1916*. G. H. von Wright & G. E. M. Anscombe (eds.). New York: Harper & Brothers.

Wittgenstein, L. (2001). *Philosophical Investigations*. Translated by G. E. M. Anscombe. Oxford: Blackwell.

Wittgenstein, L. (1974). *Tractatus logico-philosophicus*. Translated by D. F. Pears & B. F. McGuinness. London/New York: Routledge.

Wollheim, R. (1985). *Art and its objects*. Cambridge: Cambridge University Press.

Wong, P. C. & Bergeron, R. D. (1997). 30 Years of Multidimensional Multivariate Visualization. In G. M. Nielson, H. Hagen & H. Müller (eds.), *Scientific Visualization: Overviews, Methodologies, Techniques*. Los Alamitos: IEEE Computer Society, 3–33.

Wray, K. B. (2001). Science, Biases, and the Threat of Global Pessimism. *Philosophy of Science 68(3)*, 467–478.

Wyss, B. (2006). *Vom Bild zum Kunstsystem*. Köln: Verlag der Buchhandlung Walther König.

Wulf, C. & Zirfas, J. (eds.) (2005). *Ikonologie des Performativen*, München: Wilhelm Fink.

Kolibri 360 GmbH
Ferdinand-Jühlke-Straße 7
99095 Erfurt, Deutschland
produktsicherheit@kolibri360.de